FUNDAMENTALS
OF ENZYMOLOGY

FUNDAMENTALS OF ENZYMOLOGY

RATE ENHANCEMENT, SPECIFICITY, CONTROL AND APPLICATIONS

G. P. ROYER
DEPARTMENT OF BIOCHEMISTRY
THE OHIO STATE UNIVERSITY

A WILEY-INTERSCIENCE PUBLICATION
JOHN WILEY & SONS
NEW YORK   CHICHESTER   BRISBANE   TORONTO   SINGAPORE

Copyright © 1982 by John Wiley & Sons, Inc.

All rights reserved. Published simultaneously in Canada.

Reproduction or translation of any part of this work beyond that permitted by Sections 107 and 108 of the 1976 United States Copyright Act without the permission of the copyright owner is unlawful. Requests for permission or further information should be addressed to the Permissions Department, John Wiley & Sons, Inc.

Library of Congress Cataloging in Publication Data:

Royer, G. P.
  Fundamentals of enzymology.

  "A Wiley-Interscience publication."
  Includes index.
  1. Enzymes. I. Title.
QP601.R8   574.19'25   81-11359
ISBN 0-471-04675-2   AACR2

Printed in the United States of America

10 9 8 7 6 5 4 3 2 1

To Jean Royer Kohr
in memoriam

ERRATA

The diagram shown as Figure 1.10 on page 25 should be deleted. In its place, substitute the conformations shown below. Credit: Conn and Stumpf, *Outlines of Biochemistry,* 4th ed., Wiley, New York (1976), p. 101. Reproduced by permission.

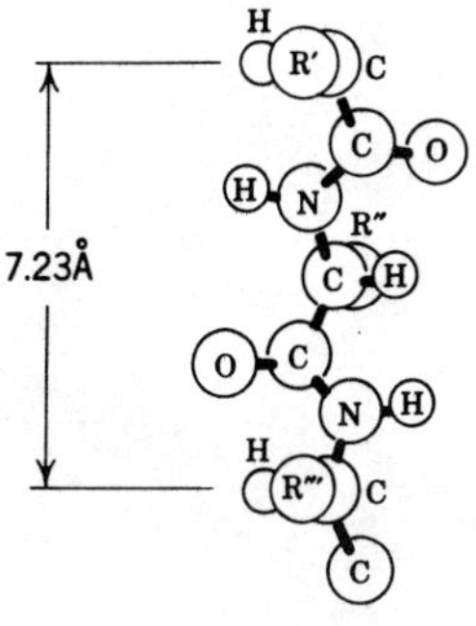

Extended chain

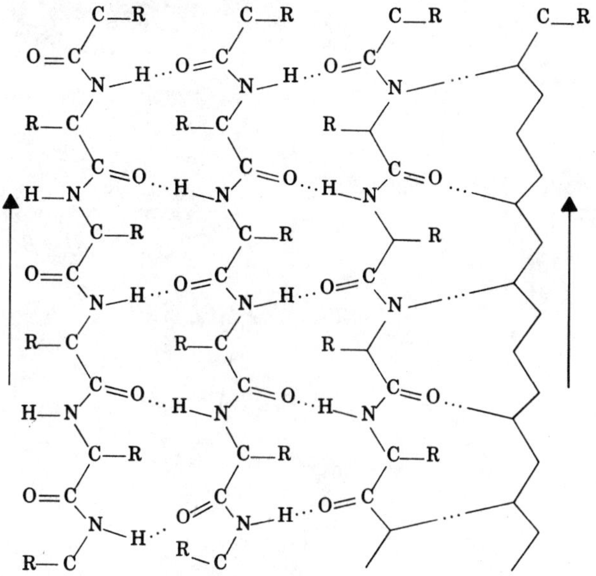

Parallel chain β-pleated sheet (stretched keratin)

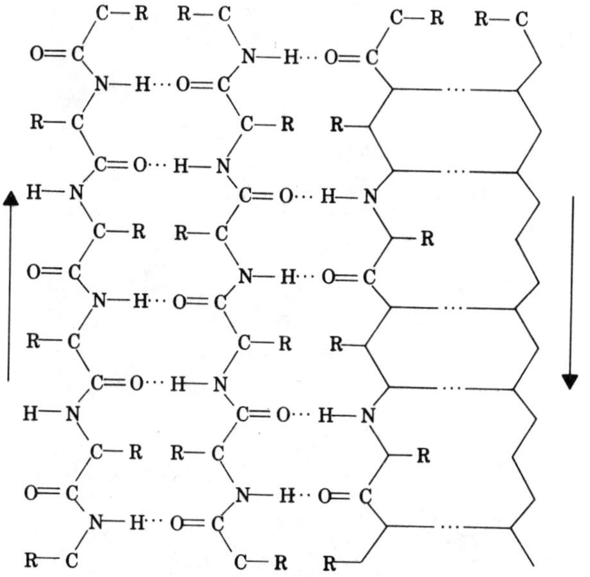

Antiparallel β-pleated sheet (silk)

PREFACE

On a number of occasions students, industrial scientists, and engineers interested in biotechnology have asked me to recommend an introductory book on enzymes that goes beyond the general biochemistry text. Although I know of comprehensive texts on enzyme mechanisms, enzyme kinetics, and control of enzyme action, I could not think of a general book that included some discussion of applications. I thought a balanced, intermediate treatise on the basics of enzyme catalysis and applications would be a useful reference.

The book is also appropriate as a text for a graduate or advanced undergraduate course. The table of contents resembles the prospectus of Biochemistry 821, a graduate course in enzymology at Ohio State University. The fundamentals are presented in the first seven weeks of the quarter; the remaining three weeks are devoted to topics from current literature that illustrate one or more basic points.

I would like to thank Jane Chapman and Doris Buchanon for help in preparation of the manuscript. Also, I acknowledge my wife, Alvilda, for her encouragement and help in proofreading.

G. P. ROYER

Worthington, Ohio
December 1981

CONTENTS

CHAPTER ONE. STRUCTURE, LOCALIZATION, AND ISOLATION   1

  A. Forces Important in the Structure and Function of Proteins, 9
      Electrostatic Interactions, 10
      The Hydrogen Bond, 16
      Van der Waals-London Dispersion Forces, 18
      Hydrophobic or Apolar Bonds, 19
  B. Enzyme Structure, 20
      Covalent Structure, 20
      Noncovalent Structure, 21
  C. Enzyme Localization, 26
  D. Enzyme Isolation, 31

CHAPTER TWO. KINETICS OF ENZYME-CATALYZED REACTIONS   39

  A. Chemical Kinetics, 40
      Order, Molecularity, and Half-Life, 40
      Transition State Theory and Catalysis, 42
  B. Rate Equations for Enzyme-Catalyzed Reactions, 43
      Henri-Michaelis-Menten Equation, 43
      Briggs-Haldane, Steady-State Approach, 45
      King-Altman Method, 47
      Reversible One-Substrate Reactions, 52
      Linear Forms of the Henri-Michaelis-Menten Equation
          for the Graphical Determination of K_m and k_{cat}, 53

- C. Effect of pH on Enzyme-Catalyzed Reaction Rates, 55
- D. Dependence of Enzyme-Catalyzed Reaction Rates on Temperature, 57
- E. Inhibition, 61
  - Noncovalent, 61
  - Covalent Inhibition, 67
- F. Bireactant Systems, 69
- G. Collection and Treatment of Enzyme Kinetics Data, 84
- H. Exercises, 88

CHAPTER THREE. STRUCTURE OF THE ACTIVE CENTER: AMINO ACID SIDE CHAINS, COENZYMES, AND METAL IONS   91

- A. Amino Acid Side Chains, 93
  - Chemical Modification with Nonspecific Reagents, 93
  - Pseudosubstrates, 94
  - Trapping of Covalent Intermediates, 96
  - Affinity Labeling, 98
  - X-Ray Crystallography, 103
- B. Coenzymes and Cofactors, 104
- C. Metal Ions, 109
- D. Conclusions, 111

CHAPTER FOUR. MECHANISMS OF ENZYME-CATALYZED REACTIONS   113

- A. General Acid-Base Catalysis, 113
- B. Nucleophilic Catalysis, 118
- C. Electrophilic Catalysis, 124
- D. Examples of Specific Enzyme Mechanisms, 127
  - Serine Proteases, 127
  - Carboxypeptidase A, 128
  - Lactate Dehydrogenase, 132
  - Lysozyme, 133
- E. Rate Enhancement in Enzymatic Reactions, 136
  - Proximity and Orientation, 136
  - Proximity of Catalytic Groups and Reaction Order, 137

Destabilization, 138
F. Conclusions, 140

CHAPTER FIVE. SPECIFICITY 143

A. Reaction Specificity, 143
B. Structural Specificity, 144
C. Stereochemical Specificity, 146
   Notation and Terminology, 146
   Examples of Stereospecificity in Enzyme-Catalyzed Reactions, 15
D. Binding and Catalytic Specificities, Lock and Key, Induced Fit, and Wrong-Way Binding, 153
E. Limitations of Enzyme Specificity, 157
F. Conclusion, 158

CHAPTER SIX. CONTROL OF ENZYME ACTION 161

A. Allosteric Enzymes, 162
B. Control by Enzyme-Catalyzed Chemical Modification (Reversible), 169
C. Control by Proteolytic Action, 30, 174
   Activation of Pancreatic Zymogens, 20, 174
   Proteolytic Activation and Blood Clotting, 175
D. Conclusion, 178

CHAPTER SEVEN. IMMOBILIZED ENZYMES 181

A. Methods of Enzyme Immobilization, 182
   Physical Methods, 182
   Chemical Methods, 183
B. Properties of Immobilized Enzymes, 186
   Physical and Chemical Modifications, 186
   The Microenvironment of Fixed Enzymes, 187
   Diffusional Effects, 190
C. Multienzyme Systems, 194
D. Applications, 196
   Large-Scale Industrial Application, 197

Use of Bound Enzymes in the Synthesis of Fine Chemicals and Pharmaceuticals, 198
Peptide Synthesis, 199
Analytical Applications, 202
Medical Applications, 202

CHAPTER EIGHT. ENZYME-LIKE SYNTHETIC CATALYSTS ("SYNZYMES"), 205

   A. Macrocycles, 206
   B. Catalysts Based on Synthetic Polymers, 210
   C. Immobilized Enzyme-Like Catalysts, 216
   D. Conclusions and Prospects, 219

INDEX 221

CHAPTER ONE

STRUCTURE, LOCALIZATION, AND ISOLATION

Enzymes are sophisticated catalysts found in all living cells. They are distinguished from catalysts of nonbiological origin by their efficacy, specificity, and sensitivity to control. An enzyme can be activated, transform a substrate* selectively at an impressive rate, and be deactivated until needed again. The catalytic power of enzymes permits biochemical reactions to go under mild conditions (37°C, 1 atm); precise specificity permits many reactions to go simultaneously with order; sophisticated control of enzymes allows the organism to respond rapidly to environmental changes.

For many years the size, complexity, and instability of enzyme molecules precluded an accurate understanding of their composition and structure. It is now well known that enzymes are proteins with molecular weights ranging from 10,000 to 500,000. Proteins are polyamides composed of the 20 coded L-α-amino acids and their derivatives, such as glycosyl and phosphoryl adducts (Table 1.1). Metal ions (Zn^{2+}, Mg^{2+}, K^+, etc.) are present in about 35% of the known enzymes. Coenzymes such as those shown in Table 1.2 are frequently present at the active centers of enzymes.

Enzymes are generally globular in shape, which means that the peptide chain

---

*"Substrate" is defined as the reactant in an enzyme-catalyzed reaction.

TABLE 1.1 The Amino Acids[a]

	Structure	pK_a's	Special Properties of Side Chains
Glycine (75)	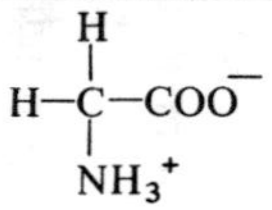 H–CH(NH$_3^+$)–COO$^-$	2.4, 9.8	Provides flexible link, often at bends in peptide chain
Serine (105)	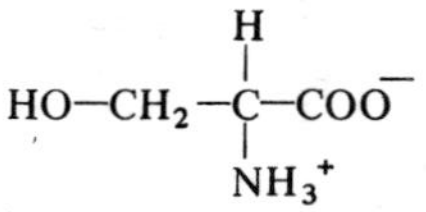 HO–CH$_2$–CH(NH$_3^+$)–COO$^-$	2.2, 9.2	Hydroxyl group is the nucleophile at the active sites of the serine proteinases Site of phosphorylation in phosphoproteins and phosphotransferases Site of glycosylation
Threonine (119)	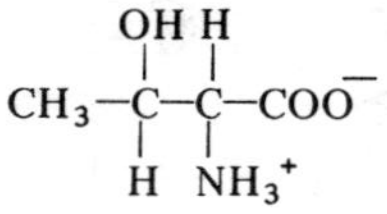 CH$_3$–CH(OH)–CH(NH$_3^+$)–COO$^-$	2.1, 9.1	Site of phosphorylation in phosphoproteins Site of glycosylation
Cysteine (121)	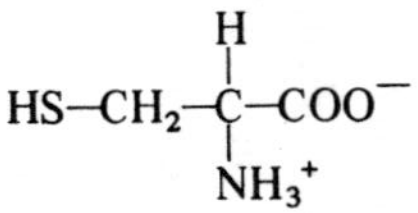 HS–CH$_2$–CH(NH$_3^+$)–COO$^-$	1.9, 10.5, 8.4 (thiol group)	Active site nucleophile in a variety of enzymes. Can be oxidized to the disulfide under mild conditions
Asparagine (132)	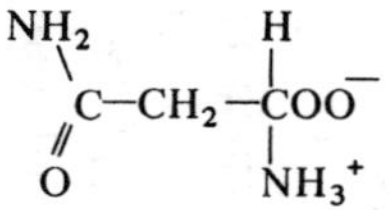 NH$_2$–C(=O)–CH$_2$–CH(NH$_3^+$)–COO$^-$	2.0, 8.8	Site of glycosylation

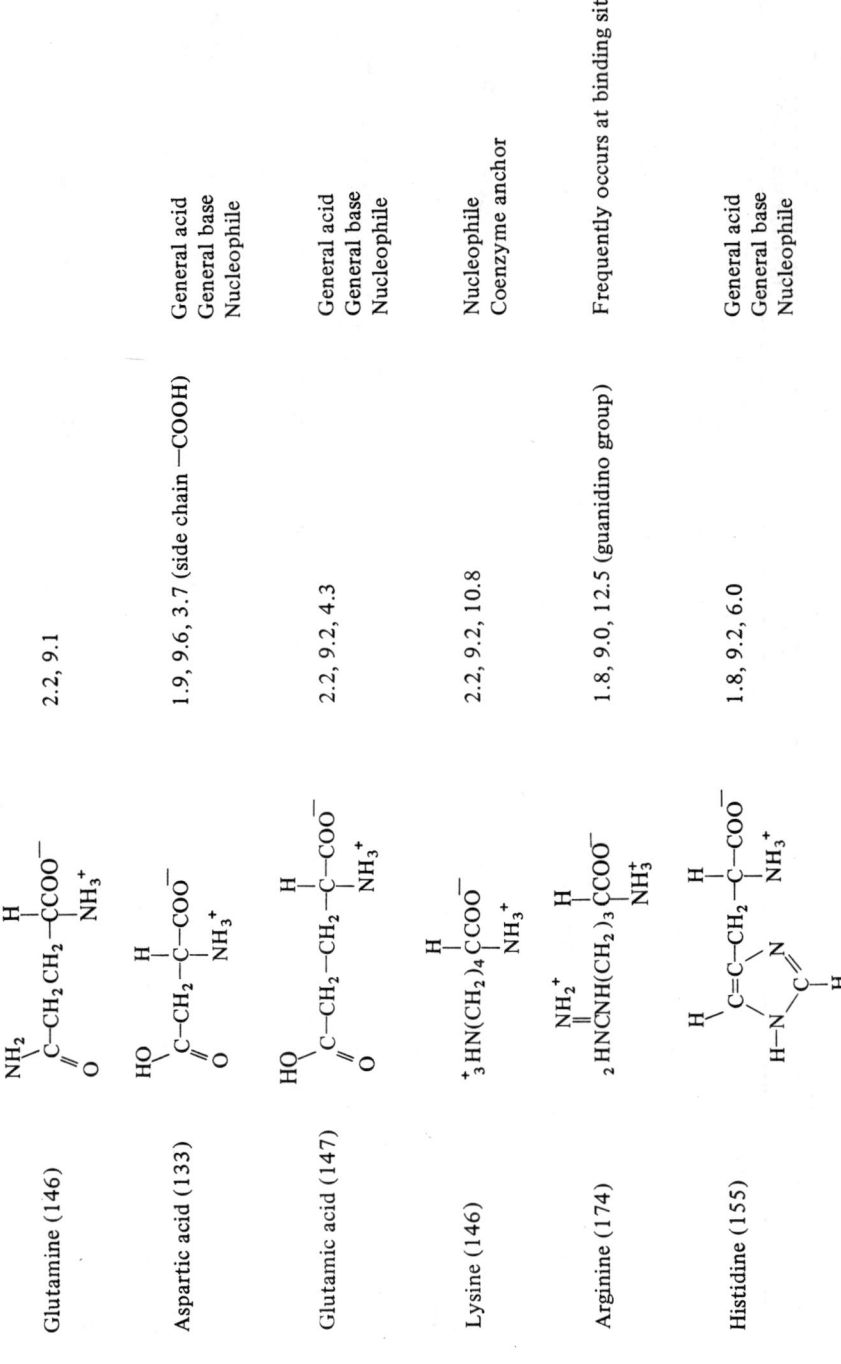

Amino acid	Structure	pKa values	Function
Glutamine (146)		2.2, 9.1	
Aspartic acid (133)		1.9, 9.6, 3.7 (side chain —COOH)	General acid / General base / Nucleophile
Glutamic acid (147)		2.2, 9.2, 4.3	General acid / General base / Nucleophile
Lysine (146)		2.2, 9.2, 10.8	Nucleophile / Coenzyme anchor
Arginine (174)		1.8, 9.0, 12.5 (guanidino group)	Frequently occurs at binding sites
Histidine (155)		1.8, 9.2, 6.0	General acid / General base / Nucleophile

TABLE 1.1. Continued.

	Structure	pK$_a$'s	Special Properties of Side Chains
Alanine (89)	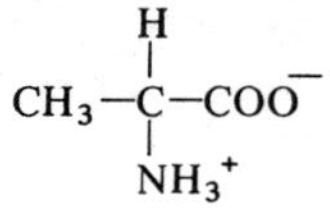 CH$_3$–C(H)(NH$_3^+$)–COO$^-$	2.3, 9.7	
Valine (117)	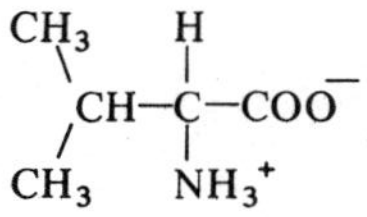 (CH$_3$)$_2$CH–C(H)(NH$_3^+$)–COO$^-$	2.3, 9.6	
Leucine (131)	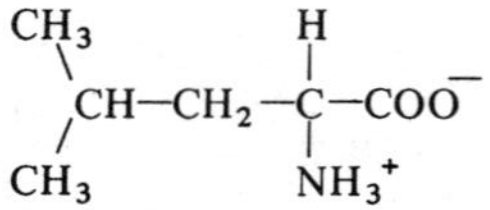 (CH$_3$)$_2$CH–CH$_2$–C(H)(NH$_3^+$)–COO$^-$	2.4, 9.6	
Tyrosine (181)	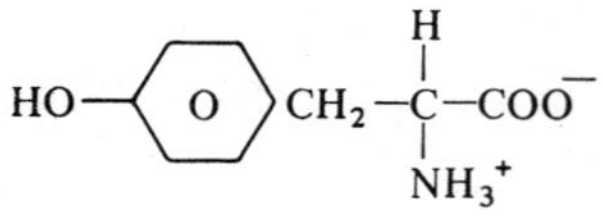 HO–C$_6$H$_4$–CH$_2$–C(H)(NH$_3^+$)–COO$^-$	2.2, 9.1, 10.1 (phenolic–OH)	Apolar General acid $E_{274.6} = 1420$ $E_{294} = 2330$ (phenolate anion)
Tryptophan (204)	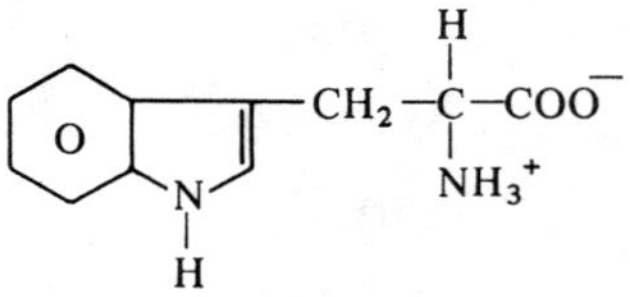 (indole)–CH$_2$–C(H)(NH$_3^+$)–COO$^-$	2.4, 9.4	UV fluorescence Apolar $E_{279.8} = 5600$ Electron donor in charge transfer complexes

Methionine (149)	CH$_3$—S—CH$_2$—CH$_2$—$\overset{\text{H}}{\underset{\underset{\text{NH}_3^+}{\mid}}{\text{C}}}$—COO$^-$	2.3, 9.2	
Isoleucine (131)	CH$_3$CH$_2$CH—CH—CO$_2^-$ $\underset{\text{CH}_3}{\mid}\underset{\text{NH}_3^+}{\mid}$	2.4, 9.7	Apolar
Proline (115)	$\begin{array}{c}\text{CH}_2\text{—CH}_2\\ \phantom{\text{CH}_2}\diagdown\diagup\text{CH}_2\text{—CO}_2^-\\ \text{CH}_2\text{N}\\ \mid\\ \text{H}\end{array}$	2.0, 10.6	Imino acid, helix breaker, frequently in sharp turns of protein chains
Phenylalanine (165)	⌬—CH$_2$$\underset{\underset{\text{NH}_3^+}{\mid}}{\text{CH}}CO_2^-$	2.2, 2.9	

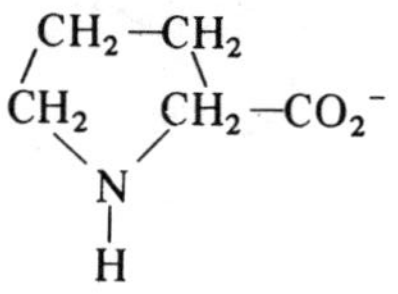

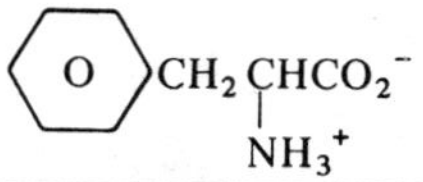

[a] pK$_a$'s from *Handbook of Biochemistry* (Herbert A. Sober, Ed.), CRC Press, Cleveland (1970).

TABLE 1.2. Coenzymes

Structure	Function
Nicotinamide Adenine Dinucleotide (NAD) 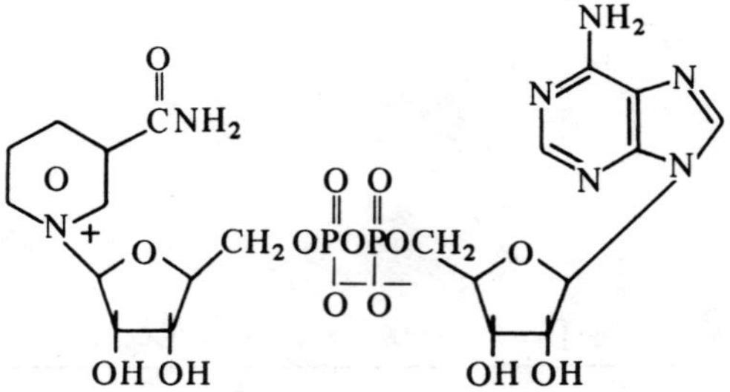	Redox Reactions
Flavin Adenine Dinucleotide (FAD) 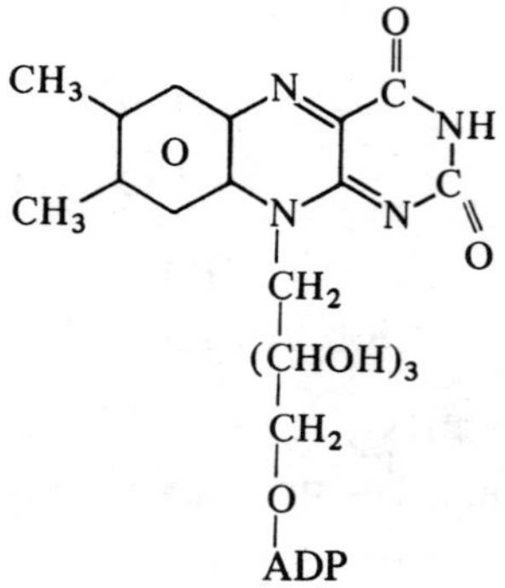	Redox Reactions

Aldehyde Group Transfer

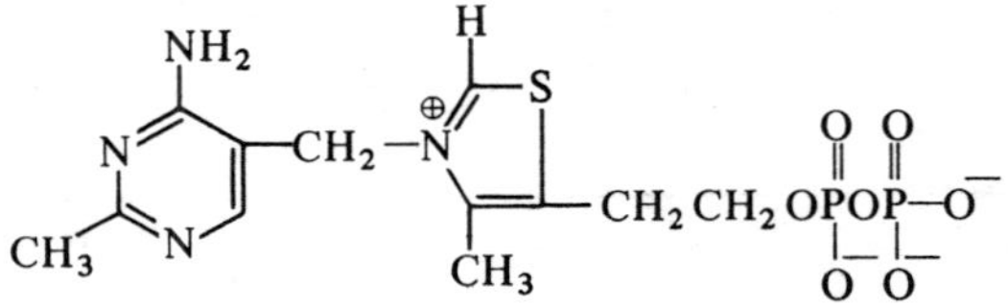

Thiamin Pyrophosphate

Amino group transfer

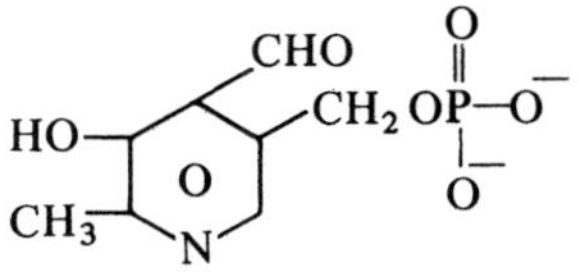

Pyridoxal Phosphate

Carboxyl Group Transfer

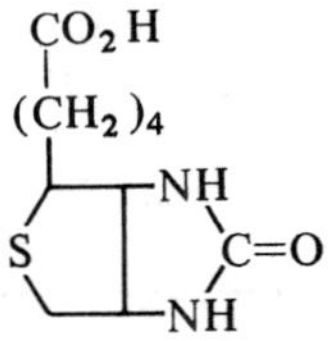

Biotin

1 Carbon Transfer

Tetrahydrofolic Acid

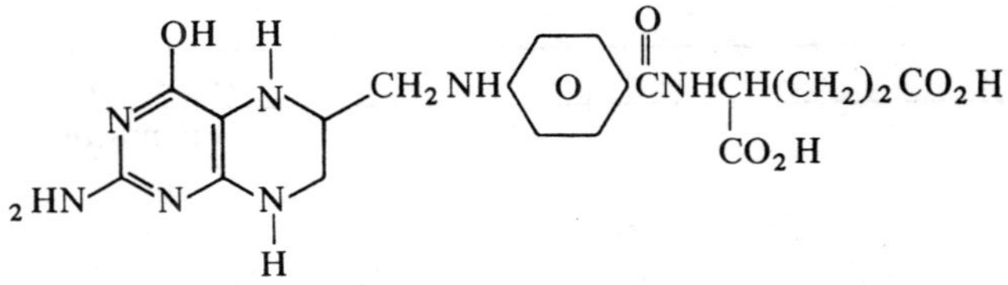

TABLE 1.2 Continued

Structure	Function
Coenzyme A 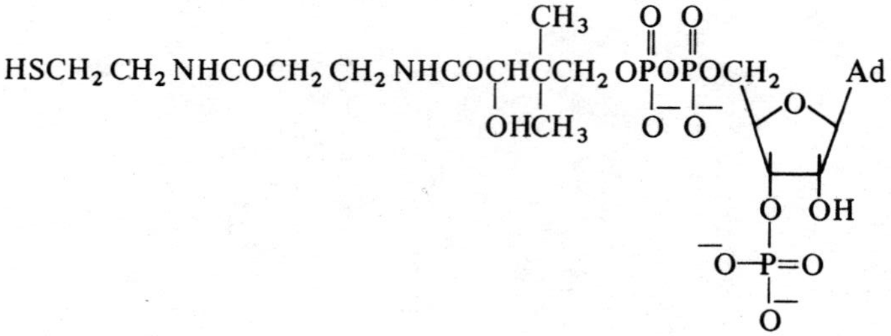	Acyl Group Transfer

$$HSCH_2CH_2NHCOCH_2CH_2NHCOCHCH_2OPOPOCH_2\text{—(ribose-Ad)}$$
with CH_3, $OHCH_3$ substituents and phosphate groups

| Ascorbic Acid 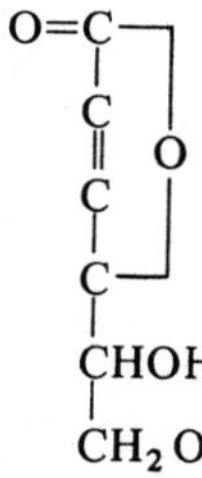 | Cofactor in Hydroxylation Reactions |

$$O=C-C=C-C-CHOH-CH_2OH$$
(with ring O)

folds back on itself repeatedly. The active center is comprised of amino acid side chains, which may be widely separated in the linear amino acid sequence. (It is known that, in ribonuclease, essential histidines at positions 12 and 119 in the linear sequence are close together at the active center.) The catalytic and substrate-binding sites constitute the active center, which represents only a small fraction of the total enzyme surface. Is the remainder of the molecule important in enzyme function? Disruption of the protein structure by physical or chemical agents (denaturation) leads to loss of control (desensitization) and activity. There is a large body of indirect evidence to suggest that substrates bring about conformational changes in the protein during the catalytic process. There is no doubt that the integrity of regulatory sites distinct from the active center (allosteric sites) is required. These areas will be discussed in detail later. At this point we can say, based on evidence concerning denaturation and the study of model compounds, that the portion of an enzyme outside the active center is important, and not simply excess baggage resulting from constraints on the evolutionary process.

A. FORCES IMPORTANT IN THE STRUCTURE AND FUNCTION OF PROTEINS

Before discussing the structure of enzymes and how they work, we shall look at the forces involved in the stabilization of protein structure and enzyme-substrate interactions. Proteins are complicated molecules and water structure is far from simple. However, some general guidelines for the consideration of forces in aqueous solution can be put forward.

1. *Two states must be considered.* One cannot predict, for instance, the solubility of a compound on the exclusive consideration of solute-solvent interaction. The interactions in the crystal must also be taken into account. In thermodynamic parlance, the relative chemical potentials of the molecules in the starting and final states must be known to predict the extent and direction of a given transformation or chemical reaction.

2. *In biopolymers a multiplicity of weak bonds can contribute significantly to the stabilization of a structure.* Van der Waals' attraction between one atom of the substrate and the enzyme may be of little consequence. However, the summation of many such weak interactions can lead to stabilization of an enzyme-substrate complex.

3. *Protein conformation and enzyme-substrate complex formation generally involve a number of forces.* Designating one type as the driving force has been a common oversimplification. One must also consider the forces responsible for specificity. Apolar bonds may provide the major driving force, but weak hydrogen bonds may provide the specificity crucial for the interaction under consideration.

4. *The values in Table 1.3 are approximations of the magnitudes of forces for stabilization of protein structure and enzyme-substrate complex formation.* Such information is difficult to obtain since enzyme-substrate interactions usually involve more than one type of force. Protein conformation, even in the simplest cases, involves all the forces listed in Table 1.3.

Electrostatic Interactions

Proteins contain many charged side chains at neutral pH. Carboxylate anions, imidazolium cations, ϵ-ammonium groups, and guanidinium cations can interact with solvent, dissolved ions, and one another. The magnitude of the electrostatic force between two charges, q_1 and q_2, is given in Eq. 1.

$$F = \frac{q_1 q_2}{Dr^2} \tag{1}$$

D is the dielectric constant and r is the distance separating the charges. The dielectric constant is a measure of the polarity of the medium. A vacuum has a dielectric constant of one. Hence D represents the electrostatic force in a given medium divided by the corresponding force in a vacuum (Eq. 2).

$$\frac{F}{F_{vac}} = \frac{q_1 q_2/Dr^2}{q_1 q_2/r^2} = D \tag{2}$$

Some dielectric constants appear in Table 1.4. The polarity within a biological cell can vary widely. The lipophilic area in membranes would have a dielectric constant below 20; that of cytoplasm would be near 100. Enzyme active centers are usually considered to be less polar than water. Bernhard gives the range of D for enzyme active centers as 10-40 (1). It has been suggested by Warshal and Levitt that the dielectric constant of the active center of lysozyme is as low as five (2). Most such estimates are made on the basis of structural information

A. FORCES IMPORTANT IN THE STRUCTURE/FUNCTION OF PROTEINS

TABLE 1.3. Approximate Magnitudes of the Forces in Aqueous Solution

Type of Bond	$\Delta G°$ (kcal/mol)
Apolar	4-8
Electrostatic	1-3
Van der Waals' interaction	<1
Hydrogen bond	<1

from X-ray diffraction studies. But differences in conformation and solvation between solution and crystal structures make these estimates rough approximations. Nevertheless, these values of D can be useful, as we shall see later.

In terms of specificity and driving force, electrostatic bonds are important in the stabilization of protein structure and the binding of substrates and cofactors. Repulsion of charged products can also be important. The formation in water of an ion pair from monovalent species is accompanied by a decrease in standard free energy of less than 1 kcal/mol (3). This number seems small, but it should be considered a basic value that is augmented by two conditions. First, "salt bridges" have been observed in proteins. This ionic bond often occurs in the protein interior in an apolar environment. The bond strength would be greater than 1 kcal/mol, as is shown in Eq. 1. The chelate effect is another factor that leads to substantial electrostatic contributions. Substrates and cofactors that carry multiple charges (e.g., PPi, ATP, NAD) are bound strongly. At least part of the strong interaction may be ascribed to the chelate effect and multiple ionic bonds.

Electrostatic contributions to substrate binding have been thoroughly studied in a number of cases. Trypsin is a proteolytic enzyme found in the small intestine of mammals. It catalyzes the hydrolytic cleavage of proteins on the carboxyl side

TABLE 1.4. Some Dielectric Constants (20°)

Substance	D
Vacuum	1.00
Hexane	1.87
Benzene	2.28
Chloroform	5.05
Ethanol	24.00
Water	80.00
Glycine in water (2.5 M)	137.00

of lysine and arginine, as well as the hydrolysis of esters, such as benzoyl-L-arginine ethyl ester (1). This reaction is strongly inhibited by benzamidine (2).

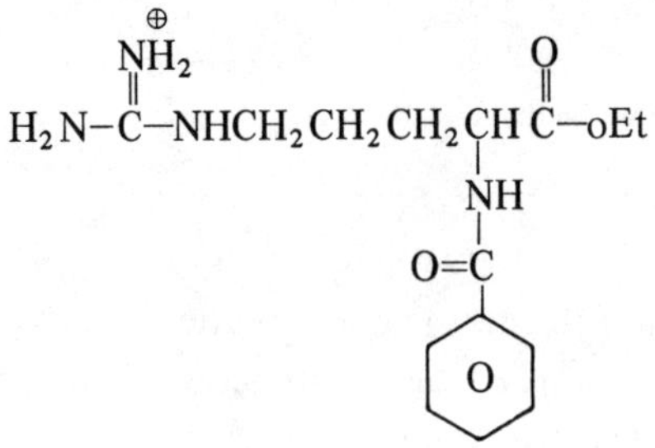

$$\underset{1}{H_2N-\overset{\overset{\oplus}{N}H_2}{\underset{\|}{C}}-NHCH_2CH_2CH_2\underset{\underset{\underset{\bigcirc}{\overset{|}{O=C}}}{\overset{|}{NH}}}{CH}\overset{O}{\underset{\|}{C}}-oEt} \qquad \underset{2}{H_2N-\overset{\overset{\oplus}{N}H_2}{\underset{\|}{C}}-NH-\hspace{-4pt}\bigcirc}$$

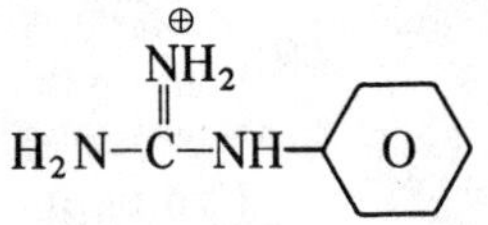

Mares-Guia and Figueriedo have studied the thermodynamics of this enzyme inhibitor and estimated the electrostatic contribution to be -2.7 kcal/mol (4). It is known that the negative charge at the trypsin binding site is contributed by the β carboxyl group of aspartic acid-177 (5,6). The carboxylate anion resides at the base of the specificity binding pocket of the enzyme (7). The significant electrostatic component of the binding free energy is probably a result of stabilization of the ionic bond by the surrounding apolar environment of the binding pocket.

Acetylcholinesterase catalyzes the hydrolysis of choline esters, such as acetyl choline (3). The reaction is inhibited by choline (4). The binding of 4 to the enzyme is 30 times stronger than the binding of the uncharged analog, isoamyl alcohol (5) (8).

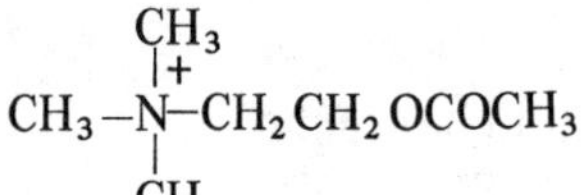

$$\underset{3}{CH_3-\overset{\overset{CH_3}{|+}}{\underset{\underset{CH_3}{|}}{N}}-CH_2CH_2OCOCH_3} \qquad \underset{4}{CH_3\overset{\overset{CH_3}{|+}}{\underset{\underset{CH_3}{|}}{N}}CH_2CH_2OH} \qquad \underset{5}{CH_3\overset{\overset{CH_3}{|}}{\underset{\underset{CH_3}{|}}{C}}CH_2CH_2OH}$$

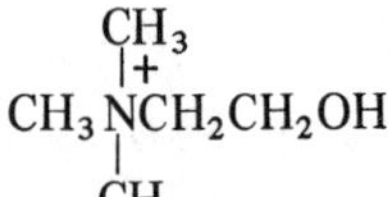

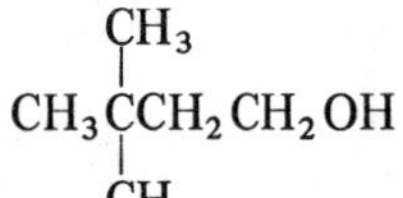

The electrostatic contribution can be calculated to be 2 kcal/mol:

$$\Delta G° = -RT \ln K; \qquad \Delta\Delta G° = -RT \ln \frac{K_4}{K_5} = -600 \ln 30 = -2040 \text{ cal}$$

A. FORCES IMPORTANT IN THE STRUCTURE/FUNCTION OF PROTEINS

Phosphate is bound to a phosphate-binding protein of *E. coli* very tightly (9). At pH 8.5 the association constant is about 10^6; at this pH value phosphate would be primarily in the diionic form. The $\Delta G°$ for binding of this doubly charged species is about −8 kcal/mol.

The three examples given above would suggest that the range of 2 to 4 kcal/mol would be an appropriate estimate for electrostatic components for ion-pair formation involved in protein-ligand interaction. Why not study enzyme-substrate interaction directly? It seems that many substrate binding constants do not reflect the true strength of the interaction. For instance, the induced-fit theory (Chapter 5) states that the binding of a specific substrate is accompanied by a conformational change of the enzyme. The apparent binding free energy would be less than the true value because of the energy required for the conformational change of the enzyme. Another difficulty is that the Michaelis constant (K_m) does not always represent a binding constant (K_s) (Chapter 2).

Classical electrostatic theory is also useful in understanding the perturbation of ionization constants at enzyme active centers. These perturbations, often several pK units, can be rationalized by dielectric effects and neighboring charge effects. The pK_a of acetic acid in water ($D = 80$) is 4.76; in 70% dioxane ($D = 18$) the pK_a is 8.32 (10). This large change in pK_a (3.6 units) may be ascribed to solvent polarity—more electrostatic work is required to bring about charge separation in an apolar medium than in a polar medium, such as water.

A neighboring charge can also produce a large perturbation in pK_a. As is shown in Table 1.5, the ionization of the carboxylic acid group is facilitated by the presence of the adjacent ammonium ion of glycine. In diglycine the effect is less because of the greater distance between the groups. The effect of intervening atoms has been shown by Bernhard for a series of diamines (Table 1.6). The perturbation is nearly zero when the distance between the amino groups becomes 11 Å. As expected from Eq. 1, the perturbation is enhanced by lowering the dielectric constant.

TABLE 1.5. pK_a Values for Selected Acids and Diacids at 25°C

Acid	pK_a
Acetic ($CH_3 COOH$)	4.76
Glycine ($NH_3^+ CH_2 COOH$)	2.35
Glycylglycine ($NH_3^+ CH_2 CONH CH_2 COOH$)	3.14

TABLE 1.6. Perturbation of pK_a's for a Series of Diamines

	r^a(Å)	ΔpK (H_2O, D^a = 80)	ΔpK (80% EtOH, D = 35)
$^+H_3N\,CH_2\,CH_2\,NH_3^+$	4.2	2.35	2.76
$^+H_3NCH_2\,CH_2\,CH_2\,CH_2\,NH_3^+$	6.6	0.90	1.25
$^+H_3N(CH_2)_8\,NH_3^+$	11.6	0.27	0.95

$^a r$ = distance between charges; D = dielectric constant.

A. FORCES IMPORTANT IN THE STRUCTURE/FUNCTION OF PROTEINS

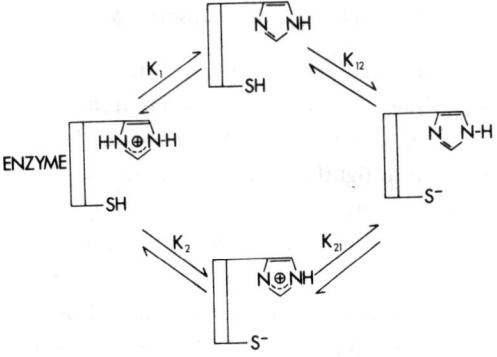

Figure 1.1. Model for the active site ionizations of papain.

Shafer and coworkers have studied the ionization of groups at the active center of papain (11). A thiol group of cysteine and an imidazole group of histidine are important for activity. On the basis of potentiometric titrations of modified and native enzyme and of NMR experiments, Shafer proposed the model shown in Figure 1.1. At 15° pK_{12} is 7.94. However, the presence of a neighboring imidazolium cation shifts this pK to 3.13 (pK_2). To test the feasibility of such a large perturbation, the dielectric constant, D, was calculated from the following equation:*

$$RT \ln \frac{K_2}{K_{12}} = \frac{q^2 N}{Dr}$$

in which q is the electronic charge, N is the Avogadro number, and r is the distance separating the imidazole group and the thiol group. From the X-ray structure, it is known that r is equal to 3.4 Å. Solving for D gives the effective dielectric constant for the area between the thiol and imidazole groups a value of 15, which is within the range given earlier.

---

*The difference in ΔG_2 and ΔG_{12} is equal to the electrostatic work required to charge the imidazole:

$$\Delta G_2 - \Delta G_{12} = -RT \ln K_2 + RT \ln K_{12} = -N \int_r^\infty \frac{q^2 dr}{Dr^2}$$

$$RT \ln \frac{K_2}{K_{12}} = \frac{q^2 N}{Dr}$$

The Hydrogen Bond

In addition to interionic attraction, dipole-dipole interaction is also important. In water the hydrogen bond, in which a hydrogen atom is shared by two other atoms (Fig. 1.2), is a good example. The donor is considered to be the atom to which the hydrogen is more tightly bound. A simple way to view this bond is as an intermediate step in the donation of a proton from an acid to a base. It is well known that water molecules form hydrogen bonds with other water molecules; one water molecule can be bonded to as many as four others. The estimates of the energy of the hydrogen bond in liquid water range from 1.3 to 4.5 kcal/mol (12). The groups in proteins that can form hydrogen bonds include $-OH, >NH$, $-NH_2$, $-COOH$ (donors), and $>C=O, >N:, -O^-, -COO^-$ (acceptors). The bond distances are between 2.6 and 3.0 Å. The strength of hydrogen bonds involving these groups is given in general biochemistry texts as 3 to 7 kcal/mol. As we shall see, this value, which may be appropriate for discussion of the solid state, has little relevance where protein structure and function are concerned. The groups listed above can form hydrogen bonds to water as well as to each other, which means *relative* hydrogen bond strengths must be considered.

Pauling, Corey, and their associates studied a number of model amides and polypeptides, using X-ray crystallography (13). Their studies revealed that maximum hydrogen bonding was obtained. Hydrogen bonds between amide hydrogen and the carbonyl oxygen are linear (within 10°), with a bond distance of 2.79 ± 0.12 Å. These studies led to the suggestion of the α-helix and β-pleated sheet structures (see below).

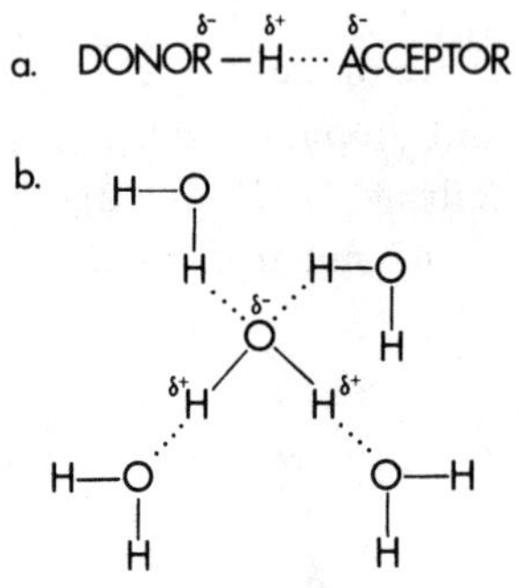

a. $\overset{\delta^-}{DONOR} - \overset{\delta^+}{H} \cdots \overset{\delta^-}{ACCEPTOR}$

b.

Figure 1.2. Hydrogen bonding in water.

Is the formation of the hydrogen bond an important driving force for the transition from random structure to helix or pleated sheet? Is hydrogen bonding important in stabilization of enzyme-substrate complexes? Using IR spectros-

A. FORCES IMPORTANT IN THE STRUCTURE/FUNCTION OF PROTEINS 17

copy, Klotz and Franzen studied the interamide hydrogen bond formation of N-methylacetamide (14).

$$2CH_3 \underset{H}{\underset{|}{C}NCH_3} \overset{O}{\overset{\|}{}} \overset{K}{\rightleftharpoons} CH_3 \underset{H}{\underset{|}{C}NCH_3} \overset{O}{\overset{\|}{}}$$
$$\vdots$$
$$CH_3 \underset{H}{\underset{|}{N}CCH_3} \overset{O}{\overset{\|}{}}$$

In water ($D = 80$) the equilibrium position for this reaction is far to the *left*. The standard enthalpy change for the reaction is zero (Table 1.7). In carbon tetrachloride ($D = 2.3$) the K is >1 and $\Delta H°$ is -4.2 kcal/mol. The entropy changes are similar for water and carbon tetrachloride systems. At this point Klotz and Franzen concluded that the stabilization of proteins by hydrogen bonds is not likely unless the bond is formed in an apolar environment. Subsequently Klotz and Farnham studied the reaction

$$>\!NH\,(aq) + O\!=\!C\!<\!(aq) \rightarrow >\!NH \cdots O\!=\!C\!<\!(org)$$

It was concluded that even this reaction is spontaneous to the left when N-methylacetamide is chosen as the model amide ($\Delta G° = +3.2$). Using data from the literature, the investigators also showed that the same can be said for the formation of the $-OH \cdots O\!=\!C\!<$ bond (Fig. 1.2c). In view of these model studies one must conclude that the process of breaking hydrogen bonds between amides and water with subsequent formation of inter-amide bonds is not conducive to the stabilization of enzyme-substrate complexes or protein conformation (14b). This does not mean that hydrogen bonds are not important in terms of specificity.

TABLE 1.7. Thermodynamics of Interamide Hydrogen Bond Formation of N-Methylacetamide at 25°C

Solvent	K	$\Delta G°$ (kcal/mol)	$\Delta H°$ (kcal/mol)	$\Delta S°$ (gibbs/mol)
CCl_4	4.7	-0.92	-4.2	-11
H_2O	0.005	3.1	0.0	-10

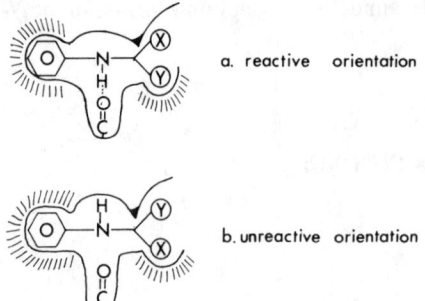

a. reactive orientation

b. unreactive orientation

Figure 1.3. Hydrogen bonding in an enzyme-substrate complex. Although the hydrogen bond does not represent a major part of the driving force of ES complex formation, it may be crucial in providing the specific reactive orientation shown in (a).

Consider the enzyme-substrate interaction shown in Figure 1.3. The apolar attractions would provide the major driving force for the interaction. In either case the hydrogen bond between the NH group and water would be broken. Formation of the enzyme-substrate hydrogen bond would make the arrangement shown in part a of the figure the preferred one over that shown in part b, and thus specify the alignment of Y and X.

Van der Waals-London Dispersion Forces

The molecules in the series shown in Figure 1.4 (methane-stannane) have no permanent dipole moments. However, these molecules show a weak attraction for each other as a result of induced dipole-induced dipole interaction. This attraction is related to the number of electrons in the molecule and is used to explain the data in Figure 1.4. The attraction changes to repulsion when electronic or-

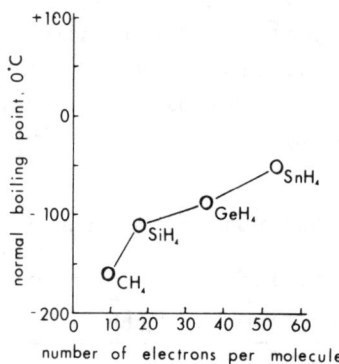

Figure 1.4. Boiling points of the homologous series methane-stannane, plotted as a function of the number of electrons to illustrate van der Waals forces.

A. FORCES IMPORTANT IN THE STRUCTURE/FUNCTION OF PROTEINS

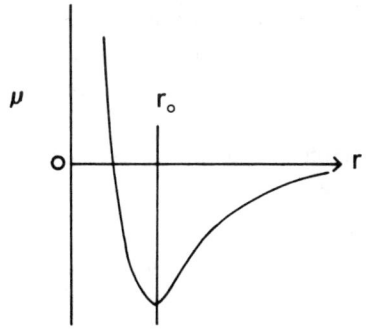

Figure 1.5. Potential energy (μ) as a function of distance (r) between atoms. The van der Waals radius is r_o.

bitals come into contact (Fig. 1.5). These forces can be described by the equation

$$u = \frac{-A}{r^6} + \frac{B}{r^{12}}$$

in which A and B are constants and r is the distance between atoms. In Figure 1.5 an energy minimum occurs at r_0, the van der Waals contact radius. Pauling points out that the sums of van der Waals radii are greater than the corresponding single-bond covalent radii by about 0.8 Å (15). Hence, it is necessary to provide energy for orbital overlap prior to covalent bond formation. In an enzyme-catalyzed reaction, this energy could come from the substrate-binding energy (see Chapter 4).

Hydrophobic or Apolar Bonds

The tendency for apolar molecules to flee from water and associate with one another is the basis of the hydrophobic bond (16). Important in both protein folding and the interaction of enzymes and antibodies with small molecules, the bond is strengthened by the presence of salt (high D) and weakened by the presence of organic solvents (low D). Kauzmann proposed that the hydrophobic bond was an important force in the stabilization of globular proteins (17). Basically, he said that the hydrophobic side groups of an extended protein chain would seek the association of other hydrophobic groups to minimize their association with water. The resulting structure would be more compact than the extended structure and would contain a core composed of hydrophobic groups. This idea was supported by early work on protein crystallography and the realization that protein denaturation by urea resulted from stabilization of the interaction between hydrophobic side chains and solvent (18). Tanford, however, has pointed out that amino acid side chains are short, and that their association requires that adjoining polar peptide segments be pulled to the interior (19). Moreover, the oil-drop model does not hold for all protein crystals (20). Nevertheless, the hydrophobic bond retains importance in the modern conception of protein folding.

Canady and coworkers have demonstrated that hydrophobic bonding is important in the formation of chymotrypsin-substrate complexes (20-25). Nonspecific hydrocarbons bind strongly to chymotrypsin and inhibit substrate binding competitively. The interaction is strengthened at high salt concentration. The $\Delta G°$ for enzyme-inhibitor complex formation varies linearly with inhibitor size (Table 1.8) for chymotrypsin and for yeast alcohol dehydrogenase. This dependence is identical to the dependence of $\Delta G°$ on inhibitor size for transfer from water to organic solvent. As a result of this common size dependence, Canady suggested that the inhibitors—and hence the apolar part of the substrate—are extracted from the aqueous phase into the enzyme phase.

TABLE 1.8. Dependence of $\Delta G°$'s on Inhibitor Size— The Basis for the Extraction Model

Transfer from H_2O to parent compound	0.12
Enzyme inhibitor complex formation	
Chymotrypsin	0.13
Yeast alcohol dehydrogenase	0.13

Hydrophobic forces are also important in the interactions of small molecules with serum proteins (26) and of antigens with antibodies (27). In Chapter 8 we we shall see several examples of model catalytic systems in which hydrophobic bonding plays an important role.

B. ENZYME STRUCTURE

Covalent Structure

As is shown in Figure 1.6, covalent structure (also called primary structure) can be represented in one dimension. It includes the amino acid sequence and arrangement of the disulfide bridges. The linear sequence of amino acids in a protein specifies the three-dimensional structure and the correct pairing of the

B. ENZYME STRUCTURE

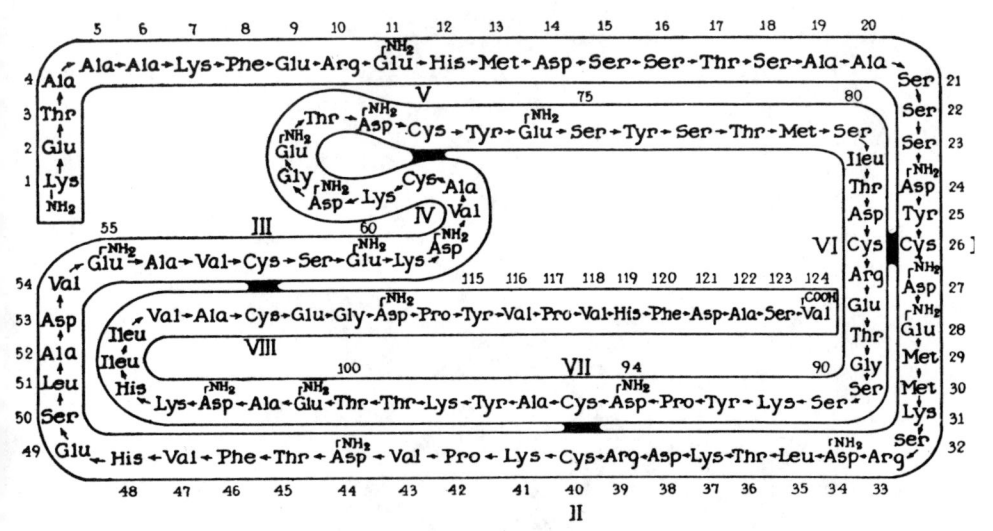

Figure 1.6. The amino acid sequence of bovine pancreatic ribonuclease A.

cysteine residues to form disulfide bridges (28). Knowledge of the sequence is essential for eventual understanding of the mechanism. The pioneering experiments of Crestfield, Stein, and Moore in the early 1960s were very important in this regard (29, 30). With the knowledge of the sequence and the results of chemical modification experiments (Chapter 3), these workers were able to show that histidyl residues at positions 12 and 119 are at the active center of ribonuclease. Knowledge of the amino acid sequence is also necessary for the construction of three-dimensional models from X-ray crystallographic data.

The peptide backbone is shown in Figure 1.7. The peptide bond has partial double bond character and may be considered as a transplanar unit. Rotations of these units occur around the C–C_α bond (Ψ angle) and the C_α–N bond (φ angle). Let us consider the number of conceivable rotational states. For a given set of Ψ, φ angles restricted to 10° intervals, there are 36^2, or 1296 states. For the protein with 100 residues there would be 1296^{100} states. Fortunately, overlap of nonbonded atoms eliminates some of the states. For instance, polyalanine of 100 residues has only about 250^{100} possible states.

Noncovalent Structure

A set of angles with high probability is $\Psi = 120°$, $\varphi = 120°$, which corresponds to the right-handed α-helix (Fig. 1.8). Other helices with dimensions that differ from those shown in Figure 1.9 include the π, 3_{10}, γ, and the ω. Another type

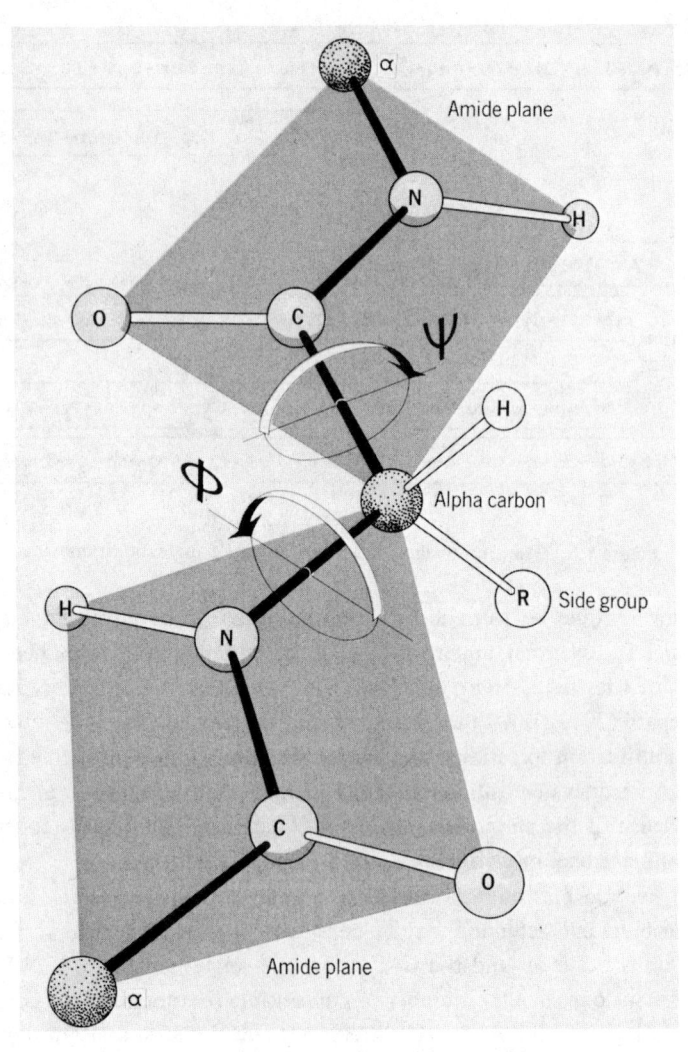

Figure 1.7. The peptide backbone, showing the transplanar peptide bond. From Conn and Stumpf, *Chemistry of Biological Compounds*, 3rd ed., Wiley, New York (1972), p. 86. Reproduced by permission.

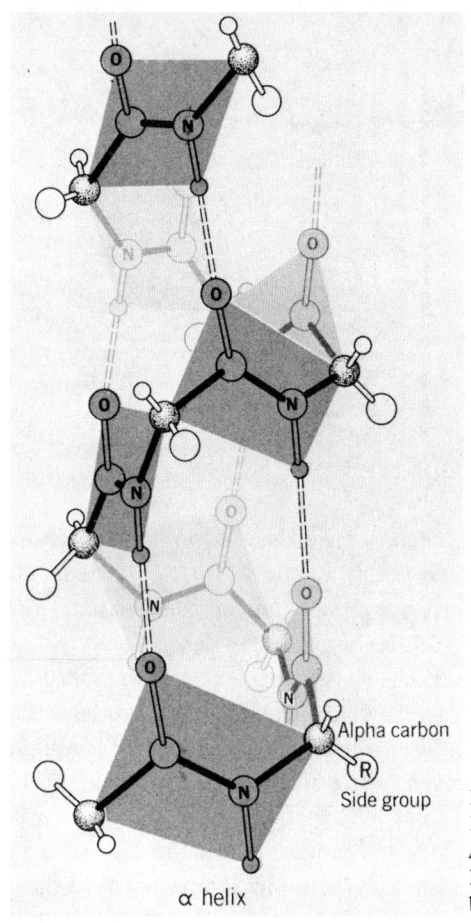

Figure 1.8. The right-handed α-helix. From Conn and Stumpf, *Chemistry of Biological Compounds*, 3rd ed., Wiley, New York (1972), p. 88. Reproduced by permission.

of regular structure is the β-pleated sheet (Fig. 1.10), in which two segments of the polypeptide line up beside one another in a way leading to formation of hydrogen bonds between the segments. The polypeptide segments can be going in the same direction (parallel β-pleated sheet) or in opposite directions (antiparallel β-pleated sheet). These regular structures, helices, and β-structures are generally classified as secondary structures.

The X-ray crystallographic structures of myoglobin and hemoglobin revealed that more than 80% of the residues were in the helical conformation. Subsequently, the structures of enzymes such as those in Figure 1.10 showed considerably less helix. Membrane proteins generally have high degrees of helicity,

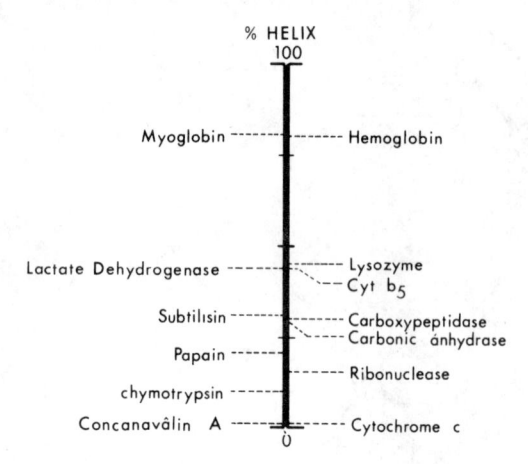

```
                        % HELIX
                         100

      Myoglobin ---------|--------- Hemoglobin



 Lactate Dehydrogenase --------|-------- Lysozyme
                                 `---- Cyt b₅
              Subtilisin --------|-------- Carboxypeptidase
                                  `----- Carbonic anhydrase
                 Papain ------|
                              |-------- Ribonuclease
           chymotrypsin ------|
           Concanavalin A ----|------- Cytochrome c
                            0
```

Figure 1.9. Helix content of some globular proteins.

presumably because of the apolar character of their environments. Short stretches of β-structure have been observed in many protein crystals. β-bends and segments of β-pleated sheet account, at least in part, for the compact nature of many enzymes.

The overall folding of the polypeptide chain is generally referred to as tertiary structure. Salt bridges, hydrophobic bonds, dipole-dipole interactions, and van der Waals' attraction provide the structure with stability (Fig. 1.11). Disruption of the structure by heat, organic solvents, and compounds such as urea and guanidine-hydrochloride lead to unfolding or denaturation. Obviously, movement of the peptide chains that contribute the groups to the active center would result in deactivation of an enzyme. In many cases denaturation is reversible. The process appears to be cooperative and relatively rapid. Metal ions, coenzymes, and prosthetic groups can facilitate refolding of the protein chain. The reader is referred to reference 31 for a comprehensive treatise on protein folding.

The last degree of protein structure is called quaternary structure and applies to the arrangement of subunits. The subunits, or protomers, may be composed of more than one polypeptide chain but are discrete functional units. Oligomers, which are collections of protomers, often contain two, four, or six identical subunits. Some oligomeric enzymes, such as aspartate transcarbamylase, contain regulatory subunits as well as catalytic subunits. When the regulatory and catalytic subunits are dissociated, the enzyme loses its sensitivity to normal regulation by small molecules (Chapter 6). A detailed discussion of the arrangement of protomers in oligomeric enzymes may be found in reference 32.

Triple helix of extended polypeptide chains typical of collagen

Figure 1.10. β-pleated sheet, parallel and anti-parallel. From Conn and Stumpf, *Chemistry of Biological Compounds*, 3rd ed., Wiley, New York (1972), p. 93. Reproduced by permission.

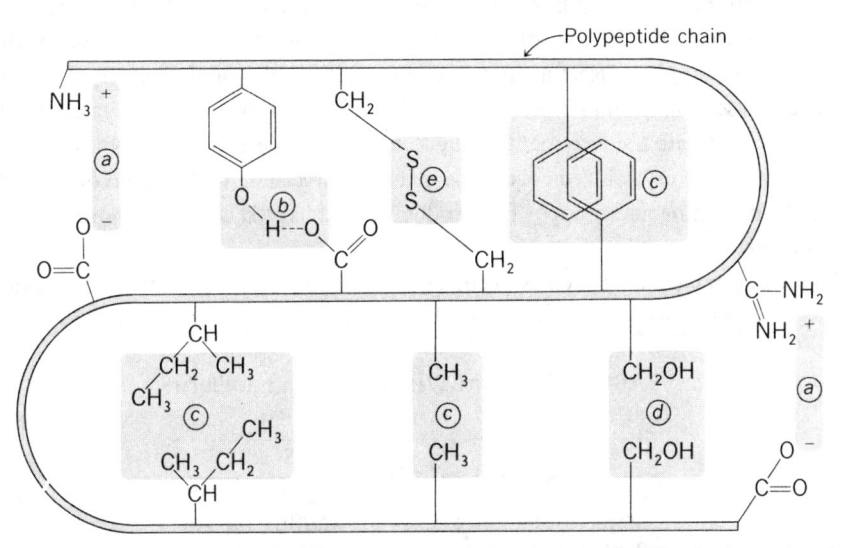

Figure 1.11. The forces that stabilize protein structure: (*a*) ion pairs, (*b*) a hydrogen bond, (*c*) apolar bonds, (*d*) van der Waals' attraction, and (*e*) a disulfide bridge.

25

To a considerable extent, enzyme structure can be correlated with the natural environment in which the enzyme works. Extracellular enzymes are generally protomeric, compact, and cross linked with multiple disulfides. Enzymes occurring within the cell (a more comfortable environment) tend to be large, flexible oligomers with few disulfides. Size and flexibility of enzymes can be related to the complexity of function and control. An intracellular enzyme may have binding sites for molecules that bear little or no resemblance to the substrates. Interaction of these molecules (effectors) with sites remote from the active center (allosteric sites) results in a conformational change and concomitant modulation of catalytic activity. Extracellular enzymes—such as microbial and pancreatic proteases, nucleases, and lysozyme—have less complicated missions and are therefore simpler in structure than intracellular enzymes.

C. ENZYME LOCALIZATION

As was stated above, enzymes can be considered in two classes—extracellular and intracellular. Extracellular enzymes tend to be more stable than intracellular, and they are almost always easier to purify. The reason for the relative ease of isolation is that the source (papaya latex, for example) contains fewer proteins than the whole cell. Some examples appear in Table 1.9. Also, cell homogenization is not required. One very interesting case for enzyme isolation from a specialized physiological fluid is that of cocoonase (33). This enzyme is produced by silkworm moths for digesting the cocoon prior to the adult's emergence. The cocoonase is made and secreted by epidermal cells in the maxillary galeae of the pupa. To isolate the enzyme, the pupae are put in a dessicator to dry the enzyme droplets. The crystallized cocoonase is then "harvested" with the aid of tweezers and a dissecting microscope. The product of this harvest is virtually homogeneous enzyme.

Most intracellular enzymes occur in one of three forms—free, fixed in multien-

TABLE 1.9. Some Well-Studied Extracellular Enzymes

Enzyme	Source
Lysozyme	Egg white
Staphyloccal nuclease	Culture filtrate
Papain	Papaya latex
Carboxypeptidase A	Pancreatic juice

zyme complexes, and bound to membranes. Relatively few free enzymes occur in the cytosol of eukaryotic cells. The enzymes of the glycolytic pathway, however, are located in cytosol along with some of the enzymes of gluconeogenesis, fatty acid synthesis, and amino acid activation. Metabolically related enzymes that occur in cytosol may be associated. Multienzyme complexes are soluble aggregates of two or more enzymes held together in a precise way by noncovalent bonds (34). There are at least two benefits of this aggregation. First, the product of the enzyme that catalyzes the first reaction of a pathway need not go far to get to the enzyme that catalyzes the second reaction. Secondly, one coenzyme molecule can be attached to the complex in such a way that it can conveniently transfer electrons or other chemical species from one enzyme to another. Both characteristics have been demonstrated in artificially immobilized enzyme systems (Chapter 7).

A number of multienzyme complexes have been characterized, and those involved in fatty acid biosynthesis, pyrimidine nucleotide biosynthesis, α-ketoglutarate oxidation, and pyruvate oxidation have been studied extensively. The pyruvate dehydrogenase complex is probably the best understood. It consists of three distinct enzymes, which catalyze the following overall reaction:

$$\text{pyruvate} + \text{NAD} + \text{CoASH} \rightarrow \text{acetyl CoA} + \text{NADH} + CO_2$$

The *E. coli* complex consists of 60 polypeptide chains and has a molecular weight of 4.6 million. There are 24 subunits of pyruvate dehydrogenase, 24 molecules of dihydrolipoyl transacetylase, and 12 molecules of dihydrolipoyl dehydrogenase. These 60 subunits associate to form a polyhedron measuring 300 Å in diameter. The transacetylase subunits are in the core, and the outside is composed of the dehydrogenases (Fig. 1.12).

The sequence of reactions catalyzed by the pyruvate dehydrogenease complex is shown in Figure 1.13. Pyruvate dehydrogenase (E_1) catalyzes the decarboxylation of pyruvate with the formation of α-hydroxyethyl-thiamine pyrophosphate

Figure 1.12. Model of the pyruvate dehydrogenase complex. From L. Stryer, *Biochemistry*, Freeman, San Francisco (1975) p. 319. Reproduced by permission.

Figure 1.13. Reactions catalyzed by the pyruvate dehydrogenase complex.

(TPP). The TPP derivative reacts with the disulfide of lipoic acid bound to the transacetylase (E_2). Thio-ester exchange with reduced coenzyme A results in the formation of acetyl CoA. The reduced lipoic acid is reoxidized by $FADH_2$ bound to the dihydrolipoyl dehydrogenase (E_3).

The functional form of lipoic acid in the complex is lipoyllysine. The lysine is contributed by the transacetylase (E_2). The eight methylene groups and the amide form a 14 Å molecular leash, which permits the lipoic acid to interact with both E_1 and E_3 (Fig. 1.14).

Enzymes are found in or on the following membranes: plasma membrane, mitochondria membranes, endoplasmic reticulum, chloroplast membrane, erythrocyte membrane, and bacterial cell wall. In transport systems the enzymes are often ATPases or "translocases." In many cases membrane-bound enzymes are related metabolically and are of the "bucket-brigade" type. An example is the family of enzymes associated with electron transport and oxidative phosphorylation located on the inner membrane of mammalian mitochondria. The reason for fixation of related enzymes on membranes is now clear. As a result of decreased diffusional paths, the lag period for the appearance of the final product

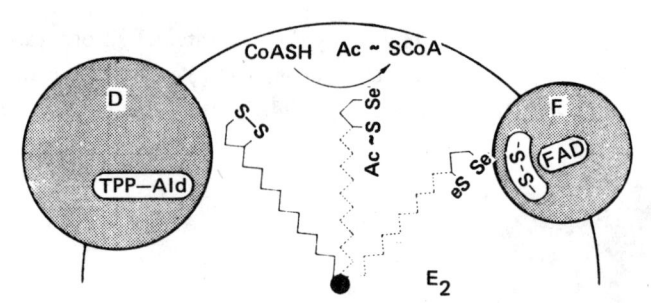

Figure 1.14. The role of lipoyllysine in the pyruvate dehydrogenase complex. See also Figure 1.13.

C. ENZYME LOCALIZATION

of a multiple-reaction sequence is significantly reduced when the enzymes that catalyze the reactions of the sequence are fixed in close proximity rather than floating freely in solution (Chapter 7). Also, channeling of metabolites from one structure or set of fixed enzymes to another is a possible point of metabolic control.

How enzymes associate with membranes is a question that occupies the attention of many current researchers. Membranes are structures consisting of proteins and amphiphilic lipids, such as cholesterol and the phosphoglycerides shown below:

CHOLESTEROL

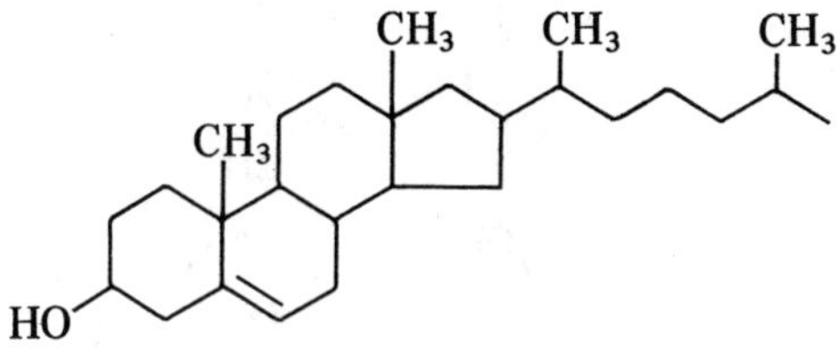

PHOSPHOGLYCERIDES

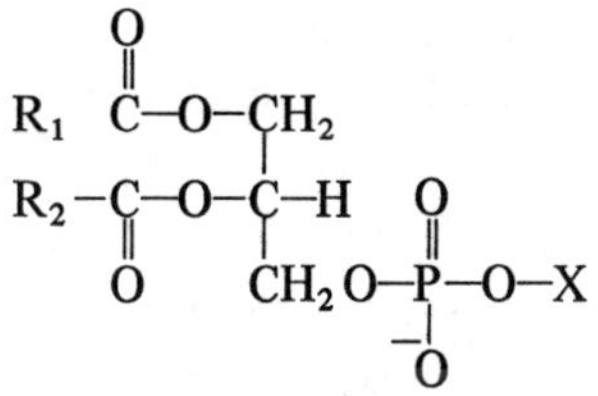

R_1 and R_2 are hydrocarbon chains containing from 16 to 24 carbon atoms. X may be neutral or positively charged at neutral pH with a variety of structures:

$-X = H$	Phosphatidic acid
$X = CH_2 CH(NH_3)COO^-$	Phosphatidyl serine
$X = -CH_2 CHOHCH_2 OH$	Phosphatidyl glycerol
$X = -CH_2 CH_2 NH_3^+$	Phosphatidyl ethanolamine
$X = -CH_2 CH_2 N(CH_3)_3^+$	Phosphatidyl choline

Cardiolipin, another important membrane lipid, is composed of two phosphatidic acid molecules joined by a glycerol molecule:

CARDIOLIPIN

$$\begin{array}{c} \text{OH} \\ | \\ \text{CH}_2-\text{C}-\text{CH}_2 \\ | | | \\ \text{O} \text{H} \text{O} \\ | | \\ ^-\text{O}-\text{P}=\text{O} \text{O}=\text{P}-\text{O}^- \\ | | \\ \text{O} \text{O} \\ | | \\ \text{CH}_2 \text{CH}_2 \\ | | \end{array}$$

Cardiolipin is also an amphiphile, which means that the molecule consists of an apolar and a polar part (Fig. 1.15). The lipid content of most membranes is 50% or less by weight. The structural models of membranes embody the lipid bilayer, which is composed of two layers of lipid with hydrocarbon tails associated. The position of protein in this bilayer has been a point of controversy. The *fluid mosaic model* of Singer and Nicholson is now the most popular (35). The proteins in this model are shown on the outside of the membrane (peripheral) or, to some extent, within the lipid bilayer (integral). The integral proteins can be partially embedded or span the bilayer completely. Peripheral proteins may be removed without disruption of the bilayer. Once removed, peripheral proteins behave like cytoplasmic or extracellular proteins in terms of solubility. Detergents must be used for the release of integral proteins.

In a very interesting study, Spatz and Strittmatter have shown that cytochrome b_5 reductase consists of a catalytic part and an additional hydrophobic segment, which is involved in the enzyme-membrane interaction (36). Such a hydrophobic anchor was also shown to exist in cytochrome b (37). These enzymes are found

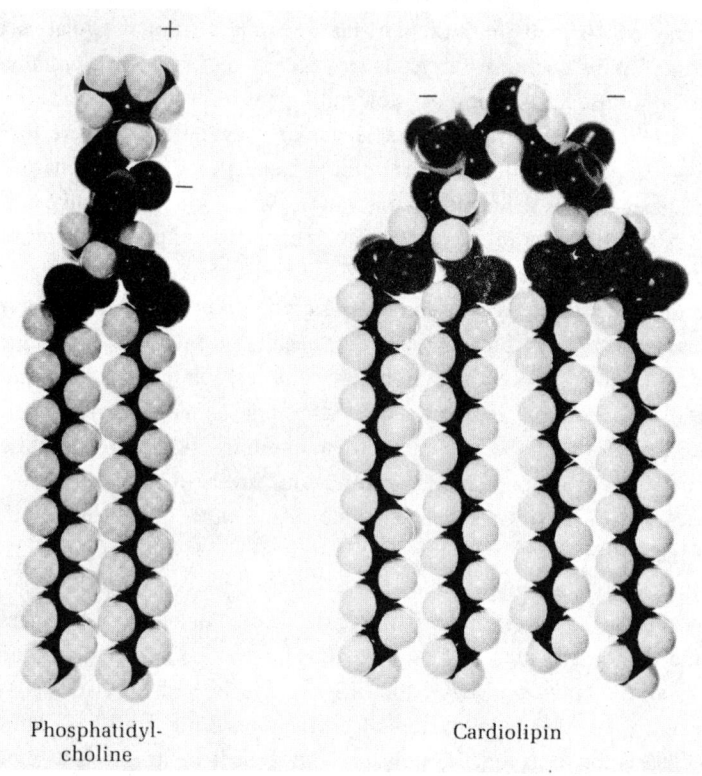

D. ENZYME ISOLATION

Phosphatidyl-
choline

Cardiolipin

Figure 1.15. A space-filling model of cardiolipin. The molecule is an amphiphile with charged phosphates at the top and long hydrocarbon chains extending downward. From Lehninger, *Biochemistry*, Worth, New York (1975), p. 289. Reproduced by permission.

in the endoplasmic reticulum and are involved in the oxidation of steroyl CoA to oleyl CoA:

D. ENZYME ISOLATION

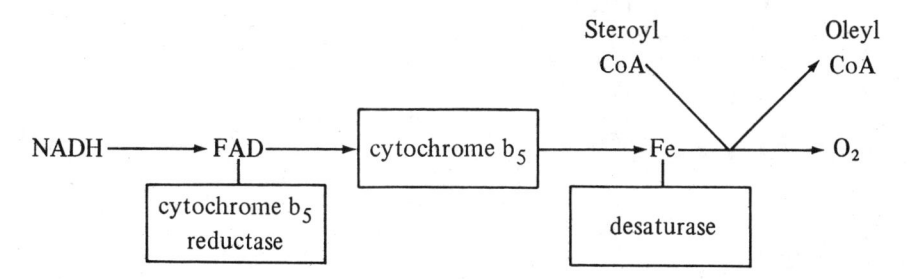

NADH ⟶ FAD ⟶ cytochrome b_5 ⟶ Fe ⟶ O_2

Steroyl CoA ↘ ↗ Oleyl CoA

cytochrome b_5 reductase

desaturase

The degree of difficulty in enzyme isolation ranges from the relative ease of "harvesting" pure cocoonase crystals to years of tedious effort, culminating in an electrophoretic purification of a few milligrams of protein. Localization is important in the isolation scheme. Extracellular enzymes in a culture filtrate are often easy to purify. Different susceptibility to heat or chemical denaturants has been used as the basis for facilitating isolation. The chymotrypsin-like enzyme of pronase is an attractive example. Pronase is a mixture of proteases from the culture filtrate of *S. griseus*. The chymotrypsin-like enzyme is stable in high concentrations of denaturant. Incubation of the crude preparation in the presence of high concentrations of urea or guanidine results in only one surviving protein, which can be purified simply by removing the fragments from the nonresistant enzymes. This example represents an exceptional case, however; generally the culture filtrate would be subjected to ammonium sulfate fractionation, ion-exchange chromatography, and gel permeation chromatography.

Purification of an intracellular enzyme begins with homogenizing the cells. Centrifugation of the cell homogenate will, in the case of eukaryotic organisms, yield the fractions shown in Table 1.10.

In most cases a detergent is required to release enzymes from membranes. Nonionic detergents, such as Triton X-100, tend to be milder than ionic detergents, such as sodium dodecyl sulfate (SDS), which binds very strongly to most proteins. Indeed, the denaturation of proteins resulting from large amounts of bound SDS is the basis for SDS-polyacrylamide electrophoresis. Lipases and proteases can also be effective for releasing membrane-bound enzymes.

The isolation of a particular enzyme is almost always a unique problem, mainly because enzymes are themselves unique structures. Let us put forth a hypothetical case. In a drug house the search is on for an enzyme to transform molecule S_A to S_B. The first step is to find a convenient assay for S_A or S_B. Coupling to a

TABLE 1.10. Fractions Arising from Progressive Centrifugation of a Cell Homogenate of Rat Liver

Centrifugation at	Fraction (in Residue)
600 g, 10 min	Nuclei and unbroken cells
15,000 g, 5 min	Mitochondria, lysosomes, and microbodies
100,000 g, 60 min	Microsomal fraction that contains the endoplasmic reticulum[a]

[a]Supernatant from this step is cytosol.

D. ENZYME ISOLATION

second reaction is common. An example is dehydrogenation (or hydrogenation) of S_B to S_C with an enzyme that uses NAD:

$$S_A \text{------} S_B \text{-------} S_C$$
$$\phantom{S_A \text{------} }NAD^+ \diagup\diagdown NADH$$

Change of absorbance at 340 nm will, in time, reflect the rate of production of S_B. The dehydrogenase must be present in a quantity sufficient to insure that dehydrogenation is very rapid compared to the first step.

After a suitable assay is found, the procedure for preparing a crude extract is optimized. Once the extract is prepared, many approaches to separations are available. References to frequently used methods for protein purification appear in Table 1.11.

TABLE 1.11. Techniques in Protein Purification

	Reference
Ammonium sulfate fractionation	38
Ion-exchange chromatography	39, 40
Gel filtration	41
Hydrophobic chromatography	42, 43
Affinity chromatography	44, 45
Electrophoresis	39, 46

Hydrophobic chromatography separates proteins on the basis of differences in the number and nature of hydrophobic amino acid side chains on the surface. A combination of hydrophobic and electrostatic forces is responsible for the interaction of some proteins with aminoalkyl-agarose (43).

Affinity chromatography is based on the biospecific interaction of an enzyme with an immobilized molecule that resembles the substrate. The first step is to select a substrate-like molecule that can be reacted with a solid support (44,45). The mixture of proteins (sometimes an initial crude extract) is applied and the column is washed (Fig. 1.16). Proteins other than the desired enzyme have no affinity for the ligand and so are washed from the column. The desired enzyme is released by elution with free ligand or a change in conditions, such as pH or temperature.

Wiedemann and Johnson have described the isolation of dihydrofolate reduc-

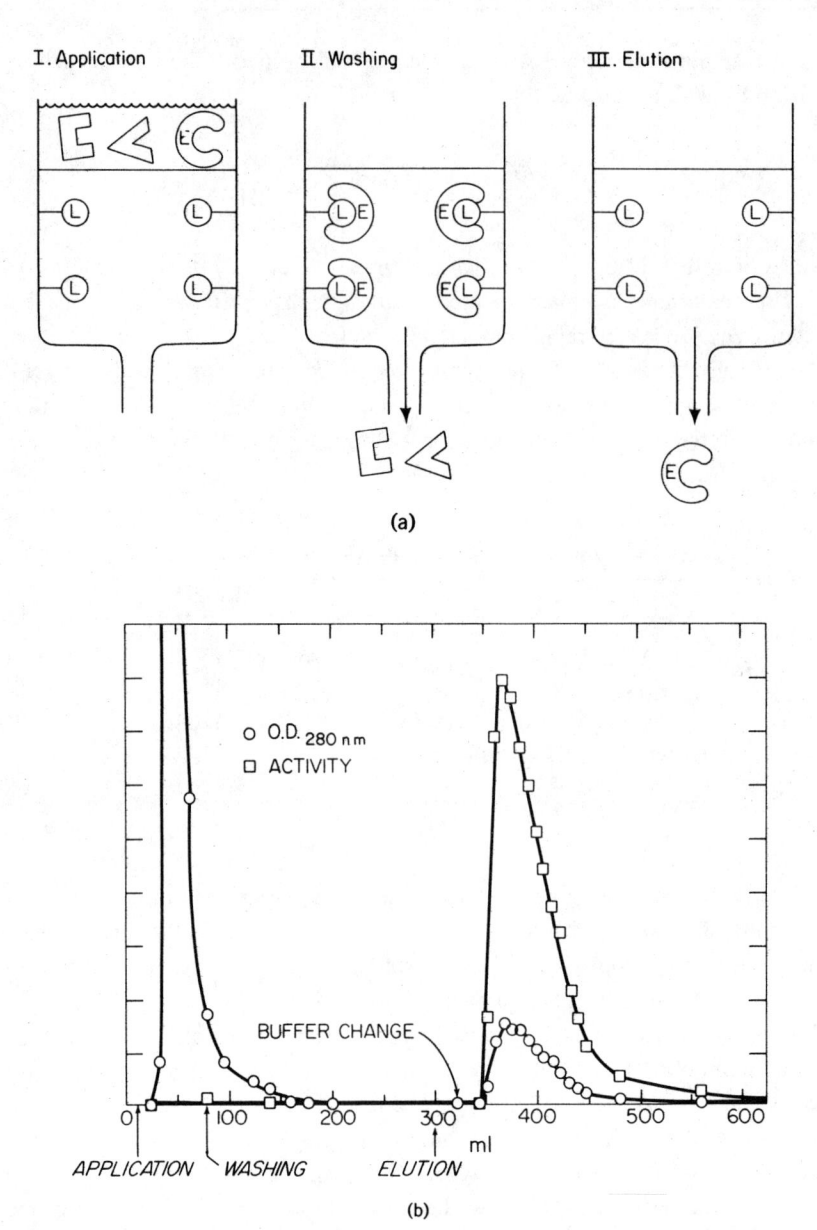

Figure 1.16. Affinity chromatography. (*a*) Schematic representation. The desired enzyme, E, binds to the immobilized ligand through a biospecific interaction. The other molecules are washed through the column. The bound enzyme can be eluted either with soluble ligand or a change of conditions (temperature, pH, etc.). (*b*) An affinity chromatogram, showing activity and protein concentration.

D. ENZYME ISOLATION

tase from an overproducing 3T6 mouse cell line (48). In Figure 1.17 the polyacrylamide gels show clearly the existence of enzyme overproduction (middle gel) and the purity of the enzyme product isolated by affinity chromatography (right gel). The affinity matrix was folate attached to agarose by means of a hexamethylenediamine anchor.

The question of what steps to employ in protein isolation is a difficult one.

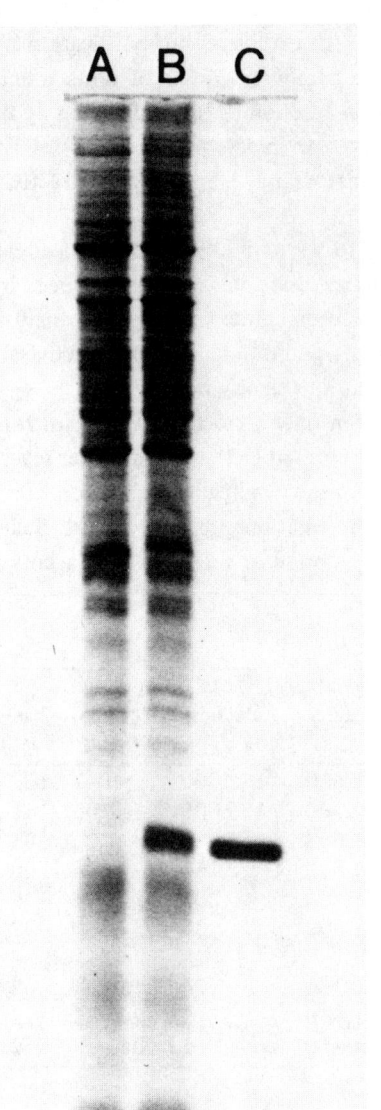

Figure 1.17. The purity of dihydrofolate reductase isolated by Wiedemann and Johnson, using a column of folate attached to agarose via an amino-hexane anchor. Gel C shows the purity of the enzyme. Gel A shows the proteins in a normal cell line, and gel B shows the overproducing line (48).

Ammonium sulfate fractionation is almost always worth a try early in the purification scheme, because volume is drastically reduced and many enzymes have good stability in the presence of ammonium sulfate. The crude ammonium sulfate fraction is a good material for application to an affinity column. A purification table, such as the one shown in Table 1.12 for renal dipeptidase, is required. An enzyme unit is generally defined as the amount of activity corresponding to the transformation of one μ mole of substrate per minute. Specific activity is the number of units per milligram of protein. In the early stages of purification it is convenient to compare one preparation to another in terms of units/volume when the total protein content is not easy to determine. The yield is simply the percentage of initial total activity at each stage. An inviolable rule is to assay and retain all fractions. Obviously, if a new chromatography step results in no activity anywhere, it should be discarded.

When has sufficient purity been obtained to warrant stopping the procedure? A single peak on gel filtration with constant specific activity across the peak is a good criterion of purity, as are bound cofactor content, end group analysis, homogeneity on acrylamide gel electrophoresis, and rational results on active site titration. Crystallization, however, does not always signify purity.

Storage conditions are important. More than once a biochemist, jubilant from a recent demonstration of homogeneity, has returned to the enzyme after several days and found it dead. Minute amounts of proteases in the preparation can be disastrous. One approach to this problem is to treat the preparation with chemical agents such as phenylmethane sulfonyl fluoride or such chelating agents as

TABLE 1.12. Purification of Renal Dipeptidase

Fraction	Volume (ml)	Total Activity (units)	Total Protein (mg)	Specific Activity X 100	Yield (%)
Kidney homogenate	3000	62,000	132,000	0.47	100
First pH 5 precipitate	2000	100,000	100,000	1.00	161
Washed precipitate	1000	66,000	55,000	1.2	106
Solubilized enzyme	1400	12,600	2,100	6.0	20
Ammonium sulfate fraction	40	5,800	200	29.0	9
Sephadex G-150	40	1,600	38.4	41.7	2.6
Carboxymethylcellulose chromatography	30	1,360	24.0	56.6	2.2
Sephadex G-200	40	1,440	6.0	240	2.3
Sephadex G-200 refiltration	40	1,410	5.8	243	2.3

EDTA (if the desired enzyme is metal-free). Cold and concentrated preparations are desirable. Freeze-drying is often useful, and glycerol can stabilize enzymes. But the freeze-thaw cycle can be harmful; freezing of small individual samples is a workable solution. An alternative is glycerol added to an enzyme solution in a hypo-vial. The nature and pH of the storage buffer are obviously important. With a new enzyme the final product should be divided into a number of lots and stored under several sets of conditions to establish the safest and most convenient storage form.

One final point of concern to those who want to set up enzyme isolation is the necessary equipment. The first requirements are a spectrophotometer, pH stat, or some other instrument for assay, and a refrigerated centrifuge and heads to handle appropriate volumes. A cold room of the walk-in variety is very convenient. Columns, peristaltic pumps, and chromatography supplies are usually basic requirements. A fraction collector, column monitor, and freeze dryer are often needed.

REFERENCES

1. S. Bernhard, *The Structure and Function of Enzymes*, Benjamin, New York (1968), p. 27.
2. A. Warshal and M. Levitt, *J. Mol. Biol.* **103**, 227 (1976).
3. W. P. Jencks, *Catalysis in Chemistry and Enzymology*, McGraw-Hill, New York (1969), p. 366.
4. M. Mares-Guia and A. F. S. Figueriedo, *Biochemistry* **9**, 3223–3227 (1970).
5. A. W. Eyl and T. Inagami, *Biochem. Biophys. Res. Commun.* **38**, 149–155 (1970).
6. A. W. Eyl and T. Inagami, *J. Biol. Chem.* **246**, 738–744 (1971).
7. R. M. Stroud, L. M. Kay, and R. E. Dickerson, *Cold Spring Harbor Symp. Quant. Biol.* **36**, 125 (1971).
8. W. P. Jencks, *Catalysis in Chemistry and Enzymology*, McGraw-Hill, New York (1969), p. 353.
9. N. Medviczky and H. Rosenberg, *Biochem. Biophys. Acta* **211**, 158 (1970).
10. J. T. Edsall and J. Wyman, *Biophysical Chemistry*, Academic Press, New York (1958), p. 472.
11. S. D. Lewis, F. A. Johnson, and J. A. Shafer, *Biochemistry* **15**, 5009 (1976).
12. D. Eisenberg and W. Kauzman, *The Structure and Properties of Water*, Oxford University Press, New York (1969), p. 179.
13. L. Pauling, *The Chemical Bond*, Cornell University Press, Ithaca, N.Y. (1967) p. 229.
14. I. M. Klotz and J. S. Franzen, *J. Am. Chem. Soc.* **84**, 3461 (1962).
14a. I. M. Klotz and S. Farnham, *Biochemistry* **11**, 3879 (1968).
15. L. Pauling, *The Chemical Bond*, Cornell University Press, Ithaca, N.Y. (1967), p. 152.

16. For a review see M. H. Klapper, *Progr. Bioorg. Chem.* **2**, 55 (1973).
17. W. Kauzman, *Adv. Prot. Chem.* **14**, 1 (1959).
18. Y. Nozaki and C. Tanford, *J. Biol. Chem.* **238**, 4074 (1963).
19. C. Tanford, *The Hydrophobic Effect*, Wiley-Interscience, New York (1973), p. 122.
20. I. M. Klotz, *Arch. Biochem. Biophys.* **138**, 704 (1970).
21. G. H. Nelson, J. L. Miles, and W. J. Canady, *Arch. Biochem. Biophys.* **96**, 545 (1962).
22. J. L. Miles, D. H. Robinson, and W. J. Canady, *J. Biol. Chem.* **238**, 2932 (1963).
23. A. J. Hymes, D. A. Robinson, and W. J. Canady, *J. Biol. Chem.* **240**, 134 (1965).
24. R. Wildnauer and W. J. Canady, *Biochemistry* **5**, 2885 (1966).
25. G. P. Royer and W. J. Canady, *Arch. Biochem. Biophys.* **124**, 530 (1968).
26. Ref. 19, p. 126.
27. F. Karush, *Adv. Immunol.* **2**, 1 (1962).
28. C. B. Anfinsen and E. Haber, *J. Biol. Chem.* **236**, 1361 (1967).
29. A. M. Crestfield, W. H. Stein, and S. Moore, *J. Biol. Chem.* **238**, 2413 (1963).
30. A. M. Crestfield, W. H. Stein, and S. Moore, *J. Biol. Chem.* **238**, 2421 (1963).
31. C. B. Anfinsen and H. A. Scheraga, *Adv. Prot. Chem.* **29**, 205 (1975).
32. I. M. Klotz, N. R. Langermann, and D. W. Darnall, *Annu. Rev. Biochem.* **39**, 25 (1970).
33. J. F. Hruska and J. H. Law, *Methods Enzymol.* **19**, 221 (1970).
34. L. A. Reed, *Accts. Chem. Res.* **7**, 40 (1974).
35. S. J. Singer and G. L. Nicolson, *Science* **175**, 720 (1972).
36. L. Spatz and P. Strittmatter, *J. Biol. Chem.* **248**, 793 (1973).
37. L. Spatz and P. Strittmatter, *Proc. Natl. Acad. Sci. USA,* **68**, 1042 (1971).
38. M. Dixon and E. C. Webb, *Enzymes*, 2nd ed., Academic Press, New York (1964).
39. Price List A, Bio-Rad Laboratories, Richmond, CA 94804 (1975), and references therein.
40. *Sephadex Ion Exchangers,* Pharmacia, Uppsala, Sweden (1979).
41. *Gel Filtration in Theory and Practice,* Pharmacia, Uppsala, Sweden (1979).
42. S. Shaltiel and Z. En-el, *Proc. Natl. Acad. Sci. USA,* **70**, 778 (1973).
43. M. Wilchek and T. Miron, *Biochem. Biophys. Res. Commun.* **72**, 125 (1976).
44. C. R. Lowe and P. D. G. Dean, *Affinity Chromatography,* Wiley, New York (1974).
45. *Methods Enzymol.* **34** (1974).
46. F. W. Studier, *J. Mol. Biol.* **79**, 237 (1973).
47. K. Hofmann, F. M. Finn, H-J. Friesen, C. Diaconescu, and H. Zahn, *Proc. Natl. Acad. Sci. USA,* **74**, 2697 (1977).
48. L. M. W. Wiedemann and L. F. Johnson, *Proc. Natl. Acad. Sci. USA,* **76**, 2818 (1979).

CHAPTER TWO

KINETICS OF ENZYME-CATALYZED REACTIONS

"You can't prove anything with kinetics."

The quotation above and others equally critical of kinetic studies have been popular in the last 15 years, probably because of the successes of the chemical and structural approaches in mechanistic studies. My response to the epigram would be, "You can't prove anything *without* kinetic studies." To do mechanistic studies without a kinetic analysis would be very difficult. Another justification for the kinetic approach is that much can be done with impure enzymes of unknown structure.

Before the development of X-ray crystallography for proteins and chemical methods for structure determination, the kinetic approach was just about the only approach available. But even today the use of kinetic methods for establishing reaction pathways is not only sensible but necessary. However, the opening quotation should be kept in mind when the temptation arises to embark on a large-scale kinetic study of the pH dependence of a complicated enzyme-catalyzed reaction in the hope of establishing the nature and behavior of groups at the enzyme active center. The conclusions drawn from such work, which often entails great investment of both human and machine time, can be not only sparse but equivocal. A balanced approach to the study of enzyme mechanism, combining kinetic and chemical methods, is usually the most productive.

This chapter deals with the very basic aspects of kinetic analysis, a subject dealt with by Segel* in a book of 957 pages! My approach in this chapter is to deal with relatively simple cases involving nonallosteric enzymes. A discussion of practical aspects of the use of kinetic studies appears at the end of the chapter, along with exercises on the derivation of rate equations.

A. CHEMICAL KINETICS

Order, Molecularity, and Half-Life

For the reaction A → P a rate equation may be written as

$$\frac{-d[A]}{dt} = k[A] \tag{1}$$

in which k is the rate constant. Simply stated, for this reaction the rate is equivalent to the product of a constant, k, and the time-dependent concentration of reactant [A]. The rate constant in Eq. 1 has the dimensions of reciprocal time and has been called the specific rate, or the rate of the reaction, when the reactant is present at unit concentration. Reactions described by the rate law in Eq. 1 are first order. In general, reaction order can be described as the sum of the powers of the concentration terms. For the reaction that obeys the rate law

$$\text{rate} = k[A][B]^2 \tag{2}$$

the overall order would be 3. The reaction may be described as first order in A and second order in B. In many cases the overall order of a reaction can be complicated by concentration terms in the denominator of the rate equation, which result from multistep mechanisms. The word "order" is not synonymous with "molecularity," which is the number of molecules participating in an elementary reaction process.

Integration of Eq. 1 gives

$$-\int_0^t \frac{d[A]}{[A]} = k \int_0^t dt$$

$$\ln \frac{A_0}{A} = kt \tag{3}$$

*This and other general books are listed at the end of the chapter.

in which A_0 is the initial concentration of reactant A. The rate law (Eq. 1) can be verified by a plot of $\ln(A_0/A)$ versus t. The plot should be linear and the rate constant, k, is the slope of the line.

The half-time, $t_{1/2}$, is the time required for the transformation of 50% of the initial amount of reactant. Thus,

$$\ln\left[\frac{A_0}{A_0/2}\right] = kt$$

$$\ln 2 = 0.693 = kt_{1/2}$$

$$t_{1/2} = \frac{0.693}{k} \tag{4}$$

The half-time or half-life of molecular processes is a useful and frequently used concept. For instance, the half-time for enzymic reactions can be in milliseconds or microseconds. The half-time for diffusional processes of small molecules is about 10^{-9} sec. The half-time for molecular bond vibration is about 10^{-13} sec.

For the reaction $2A \rightarrow P$, we may write

$$\frac{-d[A]}{dt} = k[A]^2 \tag{5}$$

For $A + B \rightarrow P$

$$\frac{-d[A]}{dt} = k[A][B] \tag{6}$$

These reactions are second order. Using the method of partial fractions, Eq. 6 may be integrated to give

$$\frac{1}{A_0 - B_0} \ln \frac{B_0[A]}{A_0[B]} = kt \tag{7}$$

To verify Eq. 6, the concentrations of A and B are determined at various times and the left side of Eq. 7 is plotted as a function of time. Second-order rate constants have the dimensions liters/mole s^{-1} or $M^{-1}s^{-1}$.

Transition State Theory and Catalysis

The concept of the activated-complex, or transition, state is very useful. For the reaction A + B → C the transition state would represents a structural intermediate somewhere between the structures of reactants or products. The transition state is described as activated because its energy level is above that of reactants or products. Figure 2.1 may be used to illustrate this idea. The reaction coordinate can be thought of as physical coordinates describing the course of the reaction in space. For the reaction H* + H-H → H*H + H the reaction coordinate is simply a distance and the transition state could be represented as H*··· H ··· H. The free energy of activation, $\Delta G^{\ddagger}$, represents the difference in free energy levels of reactants and the transition state. ΔG is the overall free energy change of the reaction. Without going into molecular considerations, it can be said that enzymes and other catalysts accelerate reactions by affecting $\Delta G^{\ddagger}$. The $\Delta G°$ and equilibrium position of the reaction are not altered by the presence of a catalyst.

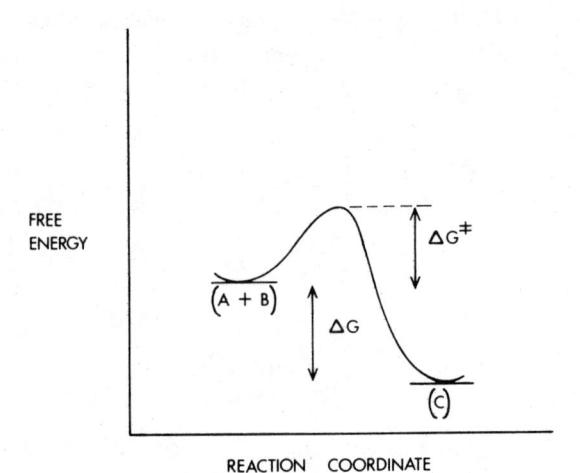

Figure 2.1. Free energy diagram showing $\Delta G^{\ddagger}$, the free energy difference between reactants A and B and the activated complex (A ··· B).

The reaction A + B → C can be written as

$$A + B \rightarrow [A \cdots B]^{\ddagger} \rightarrow C \qquad (8)$$

in which $[A \cdots B]^{\ddagger}$ is the transition state. Using this equilibrium assumption, it can be shown that

$$k = \frac{kT}{h} e^{-\Delta G^{\ddagger}/RT} \tag{9}$$

where h is Planck's constant, k is Boltzmann's constant, T is the absolute temperature, and R is the gas constant. Let us say a catalyst reduces the activation free energy, $\Delta G^{\ddagger}$, by 10 kcal/mol. The rate enhancement can then be calculated as follows:

From Eq. 9
$$\frac{k_{cat}}{k_{uncat}} = e^{(\Delta G^{\ddagger}_{uncat} - \Delta G^{\ddagger}_{cat})/RT}$$
$$= e^{(10/0.6)}$$
$$\cong 10^7$$

B. RATE EQUATIONS FOR ENZYME-CATALYZED REACTIONS

Henri-Michaelis-Menten Equation

In the early 1900s it was known that some enzyme-catalyzed reactions exhibited a hyperbolic dependence on substrate concentration. Henri and Brown proposed the idea of the formation of an enzyme-substrate complex prior to the chemical reaction step (2,3). Henri actually arrived at an equation that describes the hyperbolic substrate dependence. More than 10 years later, Michaelis and Menten published their paper, which includes a similar equation (4). A very interesting account of this and other early developments in enzyme kinetics has been prepared by Segel (5).

The reaction $S \rightarrow P$ can be illustrated as

$$E + S \underset{k_2}{\overset{k_1}{\rightleftharpoons}} ES \xrightarrow{k_3} E + P \tag{10}$$

The so-called Michaelis complex, ES, results from a noncovalent association of enzyme and substrate. One can assume that

$$\text{rate} = \frac{d[P]}{dt} = k_3 [ES] \tag{11}$$

and that

$$K_m = \frac{[E][S]}{[ES]} = \frac{k_2}{k_1} = K_s \qquad (12)$$

or that E and S are involved in a rapid equilibrium association and that the rate-determining process is the breakdown of ES to E and P. The total enzyme concentration, E_0, is the sum of free [E] and [ES].

$$E_0 = [E] + [ES] \qquad (13)$$

Combination of Eq. 12 and Eq. 13 gives

$$[ES] = \frac{E_0[S]}{K_m + [S]} \qquad (14)$$

Substitution of Eq. 14 into Eq. 11 gives a rate equation that describes the hyperbolic substrate dependence of the reaction rate

$$\frac{d[P]}{dt} = \frac{k_3 E_0[S]}{K_m + [S]} \qquad (15)$$

[S] is time-dependent. Integration is therefore necessary or a critical assumption must be made to get Eq. 15 into a form providing for the convenient determination of k_3 and K_m. The initial substrate concentration is equal to

$$S_0 = [S] + [P] + [ES] \qquad (16)$$

In studies of substrate dependence, E_0 is usually several orders of magnitude smaller than S_0. Hence.

$$S_0 \cong [S] + [P] \qquad (17)$$

At the early stages of the reaction, [P] is negligible with respect to the concentration of free substrate, or

$$S_0 = [S]$$

B. RATE EQUATIONS FOR ENZYME-CATALYZED REACTIONS

After imposing this initial velocity condition, were $(dP/dt)_0 = v_0$, Eq. 15 becomes

$$v_0 = \frac{k_3 E_0 S_0}{K_m + S_0} = \frac{k_3 E_0 S_0}{k_2/k_1 + S_0} \tag{18}$$

As we shall see later, measurement of v_0 at varying values of S_0 provides the necessary data for determination of k_3 and K_m.

Briggs-Haldane, Steady-State Approach

Rather than assume equilibrium, Briggs and Haldane used the steady-state assumption. The assumption, commonly used in chemical kinetics, is that a given intermediate is at a steady-state concentration. Stated another way, the rates of formation and removal of the intermediate are equal. Hence, $d[ES]/dt = 0$. To get an expression for [ES], Eq. 19 now replaces Eq. 12: rate of formation of ES = rate of removal of ES, or:

$$k_1 [E] [S] = k_2 [ES] + k_3 [ES]$$
$$= (k_2 + k_3) [ES]$$

or

$$[ES] = \frac{k_1 [E] [S]}{(k_2 + k_3)} \tag{19}$$

Use of Eq. 13, $E_0 = [E] + [ES]$, gives

$$[ES] = \frac{k_1 (E_0 - [ES]) [S]}{(k_2 + k_3)}$$

or

$$[ES] \frac{(k_2 + k_3)}{k_1} = E_0 [S] - [ES] [S]$$

$$[ES] [(k_2 + k_3)/k_1 + [S]] = E_0 [S]$$

$$[ES] = \frac{E_0 [S]}{(k_2 + k_3)/k_1 + [S]} \tag{20}$$

Notice the similarity between Eqs. 20 and 14. Use of Eq. 11 and the initial velocity assumption yields

$$v_0 = \frac{k_3 E_0 S_0}{(k_2 + k_3)/k_1 + S_0} = \frac{k_3 E_0 S_0}{K_{m(app)} + S_0} \qquad (21)$$

Comparison of Eq. 21 with Eq. 18 reveals that the equilibrium assumption requires that k_3 is small, relative to k_2. When K_m represents a K_s (an equilibrium constant), thermodynamic studies may be done on the enzyme-substrate interaction, using the kinetically derived binding constant K_s. Demonstration that $K_m = K_s$ can be accomplished in several ways, which are based on the observation of a change in k_3 with no change in the apparent K_m value. For a series of related substrates, if $K_{m(app)}$ remains constant and k_3 varies widely, it can be assumed that k_3 does not make a significant contribution in the expression for $K_{m(app)}$.

The steady-state assumption is less severe than the equilibrium assumption. Figure 2.2 is a time course of an enzyme-catalyzed reaction in which $E_0 \ll S_0$.

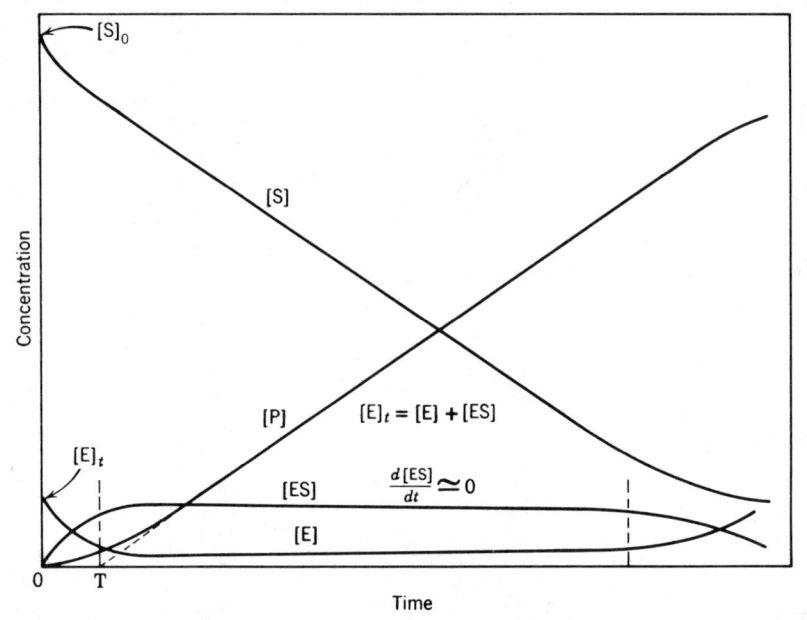

Figure 2.2 The time course of an enzyme-catalyzed reaction. The slope of the line depicting [ES] as a function of time is nearly zero over a long period after the initial burst, which is usually in milliseconds. From Segel, *Enzyme Kinetics,* Wiley, New York (1975), p. 27. Reproduced by permission.

B. RATE EQUATIONS FOR ENZYME-CATALYZED REACTIONS

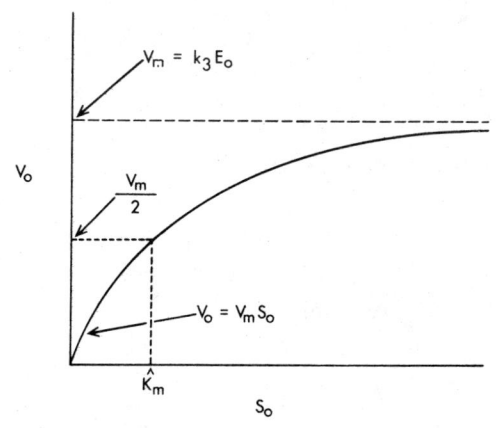

Figure 2.3. A hyperbolic dependence of V_0 on S_0 for a Michaelis-Menten enzyme system.

After an initial lag period, which is short (<1 sec), the amount of ES builds to a fairly constant level. At low levels of substrate conversion, where initial velocity studies are done, the steady-state assumption is acceptable in most cases.

Figure 2.3 is a plot of v_0 versus S_0, according to Eq. 21. When S_0 is small compared to $K_{m(app)}$, the reaction is first order in substrate concentration and the graph of v_0 versus S_0 is linear:

$$v_0 = \frac{k_3 E_0 S_0}{K_{m(app)}} ; S_0 \ll K_{m(app)} \tag{22}$$

When the enzyme is saturated with substrate and S_0 is greater than $K_{m(app)}$, the reaction is at maximum velocity and is zero order in substrate.

$$v_0 = \frac{k_3 E_0 S_0}{K_{m(app)} + S_0} = k_3 E_0 = V_m ; S_0 \gg K_{m(app)} \tag{23}$$

King-Altman Method

Many enzyme-catalyzed reactions involve more than one enzyme intermediate. In these cases the steady-state assumption can also be made, and the expressions for the concentrations of enzyme intermediates can be obtained as before. To simplify this procedure King and Altman devised a schematic method based on procedures using determinants (6). To illustrate this approach we shall consider the scheme

$$E + S \underset{k_2}{\overset{k_1}{\rightleftharpoons}} ES \overset{k_3}{\underset{P}{\searrow}} ES' \overset{k_4}{\longrightarrow} E + Q$$

which involves two enzyme intermediates. In the conventional approach,

$$E_0 = [E] + [ES] + [ES'] \qquad (24)$$

The steady-state equation for [ES] is

$$k_1 [E][S] = (k_2 + k_3)[ES] \qquad (25)$$

and for [ES']

$$k_3 [ES] = k_4 [ES'] \qquad (26)$$

The rate is given by

$$\text{rate} = k_3 [ES] \text{ (or } k_4 [ES']; \text{ see Eq. 26)} \qquad (27)$$

The expression for [ES] can be found in solving the simultaneous Eqs. 24-26 in terms of constants E_0 and S_0 (assuming we equate [S] and S_0);

$$[ES] = \frac{E_0 S_0}{(k_2 + k_3)/k_1 + [(k_3 + k_4)/k_4] S_0} \qquad (28)$$

Hence,

$$v_0 = \frac{[k_3 k_4/(k_3 + k_4)] E_0 S_0}{[(k_2 + k_3)/k_1 \cdot k_4/(k_3 + k_4)] + S_0} \qquad (29)$$

which, for comparison to Eqs. 21 and 18, may be written as

$$v_0 = \frac{k_{cat} E_0 S_0}{K_{m(app)} + S_0} \qquad (30)$$

in which

$$k_{cat} = \frac{k_3 k_4}{(k_3 + k_4)}$$

B. RATE EQUATIONS FOR ENZYME-CATALYZED REACTIONS

and

$$K_{m(app)} = [(k_2 + k_3)/k_1] \cdot [k_4/(k_3 + k_4)]$$

Let us now look at the King-Altman method applied to the same scheme. The first step is to write the scheme in cyclic form to regenerate E:

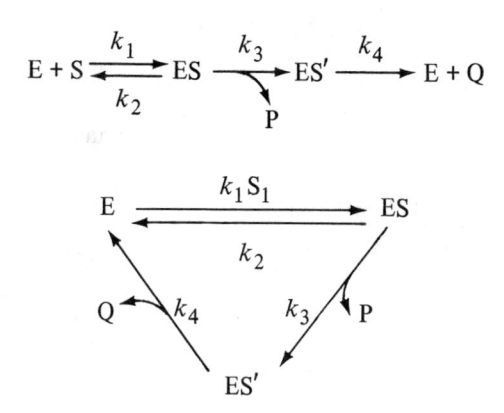

$$E + S \underset{k_2}{\overset{k_1}{\rightleftarrows}} ES \overset{k_3}{\underset{P}{\longrightarrow}} ES' \overset{k_4}{\longrightarrow} E + Q$$

E ⇌ ES (with $k_1 S_1$ forward, k_2 reverse)

Q ← (via k_4) ... ES' ← (via k_3) ... P

ES'

The figure formed is a triangle. The second step is to write down the King's patterns, which are figures with $n - 1$ lines, where n equals the number of corners of the original figure. The $n - 1$ lined patterns correspond to the generation of enzyme intermediates:

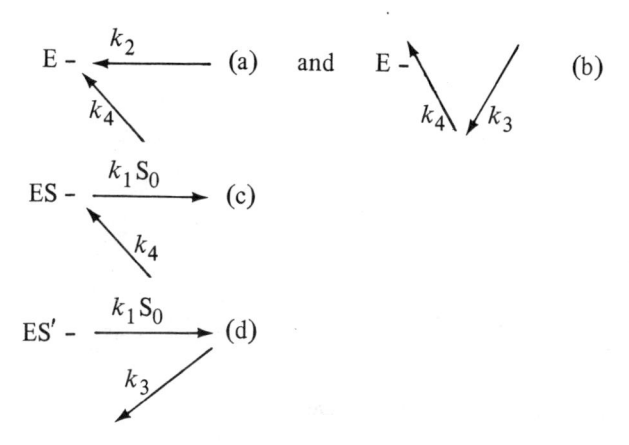

E — (with k_2, k_4) (a)   and   E — (with k_4, k_3) (b)

ES — (with $k_1 S_0$, k_4) (c)

ES' — (with $k_1 S_0$, k_3) (d)

The patterns are used to arrive at equations for the enzyme intermediates. A general distribution equation of the following type is written

$$\frac{[\text{enzyme form}]}{E_0} = \frac{\text{terms from corresponding patterns}}{\Sigma \text{ all terms}} \tag{31}$$

For free enzyme

$$\frac{E}{E_0} = \frac{k_2 k_4 + k_3 k_4}{\Sigma \text{ all terms}} \tag{32}$$

Each $n - 1$ lined figure provides a product term in the numerator: for ES,

$$\frac{[ES]}{E_0} = \frac{k_1 k_4 S_0}{\Sigma \text{ all terms}} \tag{33}$$

for ES',

$$\frac{[ES']}{E_0} = \frac{k_1 k_3 S_0}{\Sigma \text{ all terms}} \tag{34}$$

Returning to Eq. 27,

$$v_0 = k_3 [ES] \tag{27}$$

From Eq. 33,

$$[ES] = \frac{k_1 k_4 S_0}{\Sigma \text{ all terms}} \tag{34}$$

or

$$[ES] = \frac{k_1 k_4 E_0 S_0}{k_2 k_4 + k_3 k_4 + (k_1 k_3 + k_1 k_4) S_0} \tag{35}$$

Combination of Eq. 27 and Eq. 35 gives

$$v_0 = \frac{k_1 k_3 k_4 E_0 S_0}{k_2 k_4 + k_3 k_4 + (k_1 k_3 + k_1 k_4) S_0} \tag{36}$$

B. RATE EQUATIONS FOR ENZYME-CATALYZED REACTIONS

Rearrangement of Eq. 36 gives Eq. 29

$$v_0 = \frac{k_4 k_4/(k_3 + k_4) E_0 S_0}{[(k_2 + k_3)/k_1 \cdot k_4/(k_3 + k_4)] + S_0} \quad (29)$$

which was dervied from the conventional approach.

For more complicated schemes it is helpful to know how many $n - 1$ lined patterns are possible. This number is given by Eq. 37

$$\text{Total } n - 1 \text{ lined patterns} = \frac{m!}{(n-1)!\,(m-n+1)!} \quad (37)$$

Again, n equals the number of corners and m is the number of lines in the complete figure. For our scheme (a triangle) $n = m = 3$. Hence,

$$\text{Total } n - 1 \text{ lined patterns} = \frac{1 \cdot 2 \cdot 3}{(1 \cdot 2)(1)} = 3$$

There are (a) and (c) above, (b) and (d). It is possible that an $n - 1$ lined pattern is a closed loop. This is not permissible and is subtracted. Consider the following figure

in which $n = 5$ and $m = 6$. The total number of $n - 1$ lined patterns is 15, from Eq. 37. Three of the four-sided figures are closed loops, which leaves 12 possible patterns:

Closed loops (not allowed)

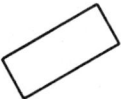

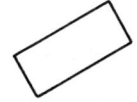

Permissible four-sided patterns

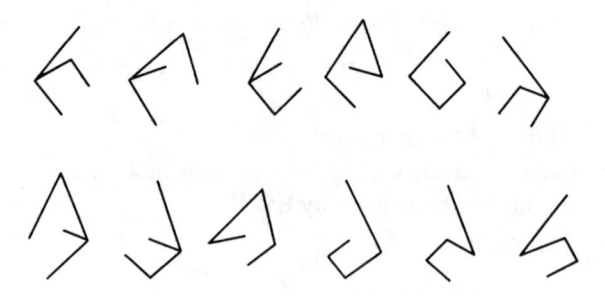

Reversible One-Substrate Reactions

The reactions written above contain irreversible steps leading to product formation. The reversible cases are quite common, however. Consider the following reaction:

$$E + S \underset{k_2}{\overset{k_1}{\rightleftharpoons}} ES \underset{k_4}{\overset{k_3}{\rightleftharpoons}} E + P \qquad (38)$$

The net rate equation from left to right is

$$v_{net} = k_3 [ES] - k_4 [E][P] \qquad (39)$$

The steady-state equation for [ES], which is derived as before, may be combined with the enzyme conservation equation and Eq. 39 to yield the rate equation from left to right:

$$v_{net} = \frac{V_{max,f}[S]/K_{m,s} - V_{max,r}[P]/K_{m,p}}{1 + [S]/K_{m,s} + [P]/K_{m,p}}$$

in which

$$V_{max,f} = k_3 E_0, \quad V_{max,r} = k_2 E_0 \qquad (40)$$

$$K_{m,s} = \frac{k_2 + k_3}{k_1}, \quad K_{m,p} = \frac{k_2 + k_3}{k_4}$$

B. RATE EQUATIONS FOR ENZYME-CATALYZED REACTIONS

In the absence of P, Eq. 39 reduces to the standard Michaelis-Menten equation for the forward direction. In the case of [S] = 0, the Michaelis-Menten equation for the reverse reaction obtains. Haldane showed the relationships between kinetic constants and the equilibrium constant. For reaction 38 the equilibrium constant for the overall reaction is given by

$$K_{eq} = \frac{[P]_{eq}}{[S]_{eq}} \tag{41}$$

If one assumes a rapid equilibrium for the enzymatic steps, by necessity

$$K_{eq} = \frac{k_1}{k_2} \cdot \frac{k_3}{k_4} \tag{42}$$

Combination of Eq. 42 and Eq. 40 gives the Haldane equation.

$$\frac{V_{max,f} K_{m,p}}{V_{max,r} K_{m,s}} = K_{eq} \tag{43}$$

Linear Forms of the Henri-Michaelis-Menten Equation for the Graphical Determination of K_m and k_{cat}

Knowledge of K_m and k_{cat} is essential for enzyme characterization and eventual understanding of mechanism. The most widely used method for determining these values is that of Lineweaver and Burk (7). The general hyperbolic rate equation can be put in linear form:

$$v_0 = \frac{V_m S_0}{K_{m(app)} + S_0} \quad ; V_m = k_{cat} E_0 \tag{44}$$

Inversion gives

$$\frac{1}{v_0} = \frac{K_{m(app)}}{V_m} \cdot \frac{1}{S_0} + \frac{1}{V_m} \tag{45}$$

The Lineweaver-Burk plot ($1/v_0$ vs $1/S_0$) is shown in Figure 2.4. When $1/v_0 = 0$,

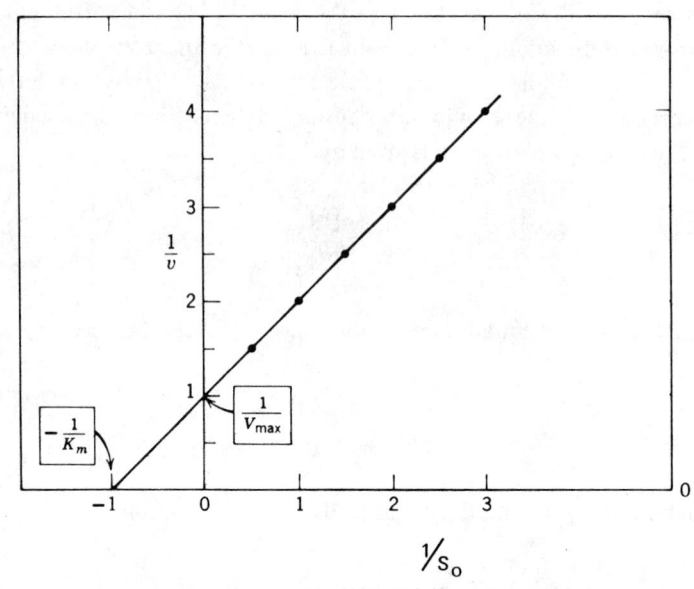

Figure 2.4. Lineweaver-Burk plot, $1/V_0$ versus $1/S_0$ also called the double reciprocal plot. From Segel, *Enzyme Kinetics*, Wiley, New York (1975), p. 47. Reproduced by permission.

$1/S_0 = -1/K_m$, and at $1/S_0 = 0$, $1/v_0 = 1/V_m$. The slope of the plot is equal to $K_{m(app)}/V_m$. (How to get the apparent constants K_m and k_{cat} in actual practice is discussed at the end of this chapter.)

An alternative plot was suggested by Eadie (8). Eq. 44 can be rearranged to

$$v_0 K_{m(app)} + v_0 S_0 = V_m S_0 \tag{46}$$

which can be written as

$$\frac{v_0}{S_0} = \frac{-v_0}{K_{m(app)}} + \frac{V_m}{K_m} \tag{47}$$

Hence, a plot of v_0/S_0 versus v_0 will give a straight line with slope equal to $-1/K_{m(app)}$ (Fig. 2.5). When $v_0/S_0 = 0$, $v_0 = V_m$, and at $v_0 = 0$, $v_0/S_0 = V_m/K_m$.

C. EFFECT OF pH ON ENZYME-CATALYZED REACTION RATES

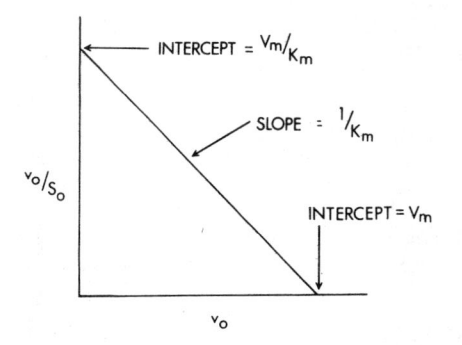

Figure 2.5. The Eadie-Hofstee plot, v_0/S_0 versus v_0.

C. EFFECT OF pH ON ENZYME-CATALYZED REACTION RATES

The influence of [H⁺] on enzyme-catalyzed reactions is generally quite complex. The proton is a simple chemical species, but the groups in proteins (and substrates) that bind protons are numerous and diverse in structure. Another complication is that ionizable groups in enzymes and enzyme-substrate complexes can exist in a multitude of microenvironments. Proton binding can affect enzyme-substrate interactions, the association state of free enzyme, conformation, chemical changes in ES, and finally, the stability of the enzyme. You can easily imagine the complexity of reaction schemes that embody these effects. The rate expressions will, of course, reflect the complexity of these schemes. We shall examine a very simple case.

A bell-shaped curve, which represents the rate of an enzyme-catalyzed reaction plotted as a function of pH, appears in Figure 2.6. The curve can be simulated with an equation based on the following reaction sequence:

$$EH_2 \underset{}{\overset{K_1}{\rightleftarrows}} EH \underset{}{\overset{K_2}{\rightleftarrows}} E$$

$$EH + S \underset{k_2}{\overset{k_1}{\rightleftarrows}} EHS \overset{k_3}{\longrightarrow} EH + P$$

KINETICS OF ENZYME-CATALYZED REACTIONS

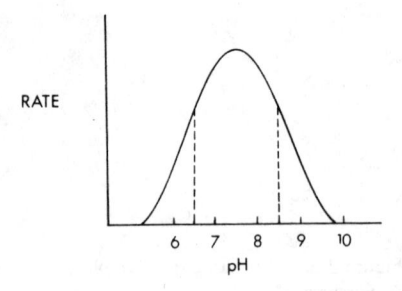

Figure 2.6. The bell-shaped plot of the rate of an enzyme-catalyzed reaction rate versus pH.

For the data in the pH rate profile shown in Figure 2.6, a schematic picture could be represented as

$$\text{Im-H}^+ \text{ NH}_3^+ \underset{\text{enzyme}}{\overset{K_1}{\rightleftarrows}} \text{Im } \text{NH}_3^+ \underset{\text{enzyme}}{\overset{K_2}{\rightleftarrows}} \text{Im } \text{NH}_2 \underset{\text{enzyme}}{}$$

$$(EH_2) \qquad\qquad (EH) \qquad\qquad (E)$$

Active form

in which Im represents the imidazole group of histidine ($pK_a \cong 6.5$) and NH_2 the α- or ϵ-amino group ($pK_a \cong 8.5$). Assuming a steady state in EHS and rapid equilibria of the proton dissociation steps, the following rate expression results:

$$v_0 = \frac{k_3 E_0 \, S_0/K_m}{(1 + K_2/[\text{H}^+] + [\text{H}^+]/K_1) + S_0/K_m} \tag{48}$$

From the inflection points in Figure 2.6, it can be seen that pK_1 and pK_2 differ by about two units, which means a factor of 100 in K. Hence, $K_1 \gg K_2$. This means that the function $(1 + K_2/[\text{H}^+] + [\text{H}^+]/K_1)$ goes through a minimum as $[\text{H}^+]$ is varied. It follows that v_0 will go through a maximum as $[\text{H}^+]$ is varied. The first derivative of the above function with respect to $[\text{H}^+]$ is:

$$F([\text{H}^+]) = \frac{1 + K_2}{[\text{H}^+]} + \frac{[\text{H}^+]}{K_1} \tag{49}$$

$$\frac{dF([\text{H}^+])}{d[\text{H}^+]} = \frac{-K_2}{[\text{H}^+]^2} + \frac{1}{K_1} \tag{50}$$

This derivative is equal to zero at the pH optimum. Using this condition, Eq. 50 becomes

$$0 = \frac{-K_2}{[H^+]_{opt}^2} + \frac{1}{K_1}$$

or

$$[H^+]_{opt} = (K_1 K_2)^{1/2} \tag{51}$$

Eq. 51 can be used to check the dissociation constants, which are derived from replots based on equations arising from Eq. 48 or by computer-assisted curve fitting (9). The apparent pK values may be taken directly from the midpoints of the ascending and descending halves of a graph such as Figure 2.6. The major problem arises in deriving mechanistic information from the values. Assignment of pK values to side chain groups is difficult because of perturbation by local environments. Heats of ionization can sometimes be helpful, but the matching of pK_a's and functional groups in proteins is often complicated and speculative. However, pK_a values and heats of ionization can be very helpful when used to confirm results of chemical modification studies.

D. DEPENDENCE OF ENZYME-CATALYZED REACTION RATES ON TEMPERATURE

The graph in Figure 2.7 shows a typical temperature dependence of the rate of an enzyme-catalyzed reaction. The initial increase in rate is followed by a precipitous drop at about 60°C. There are three basic temperature effects (Eq. 49): (1) the temperature dependence of K_m (substrate binding in the simplest case), (2) the temperature dependence of k_3, and (3) thermal denaturation of the enzyme (a decrease in E_0).

$$E + S \underset{K_m}{\rightleftharpoons} ES \xrightarrow{k_3} E + P \tag{49}$$
$$\updownarrow$$
$$E^* \text{ (inactive)}$$

The effect of temperature on K_m can be either positive or negative—enzyme-

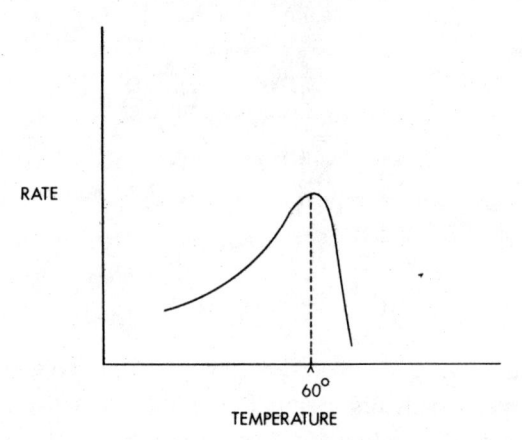

Figure 2.7. The temperature dependence of the rate of an enzyme-catalyzed reaction with a representative T_{opt} of 60°C.

substrate interaction can be endothermic or exothermic. When $K_m = K_s$, thermodynamic parameters can be determined in ways analogous to uncatalyzed systems. K_m can be determined at various temperatures. A plot of $\ln K_m$ versus $1/T$ will yield the standard enthalpy change, $\Delta H°$; according to the van't Hoff equation

$$d \ln K_m / d(1/T) = -\Delta H°/R \qquad (50)$$

the slope of the line may be positive or negative. Having determined $\Delta H°$, the standard entropy change, $\Delta S°$, can be calculated:

$$\Delta G° = -RT \ln K_m = \Delta H° - T\Delta S°$$

$$\Delta S° = R \ln K_m + \frac{\Delta H°}{T} \qquad (51)$$

The magnitude and signs of these thermodynamic parameters can be useful in the study of the enzyme-substrate interactions.

The temperature dependence of k_3 does not differ from that for rate constants in uncatalyzed systems. The Arrhenius law describes the temperature dependence of rate constants:

D. DEPENDENCE OF ENZYME-CATALYZED REACTION RATES

$$k_3 = Ae^{-\Delta E/RT}$$

or

$$\ln k_3 = \ln A - \frac{\Delta E}{RT} \quad \text{(a)}$$

$$\frac{d \ln k_3}{d(1/T)} = \frac{-\Delta E}{R} \quad \text{(b)}$$

(52)

in which A is a constant and ΔE is the experimental activation energy. A plot of $\ln k_3$ versus $1/T$ will be linear, with a slope of $-\Delta E$. Since ΔE is always positive, the Arrhenius plot should have a negative slope, as shown in Figure 2.8. The ex-

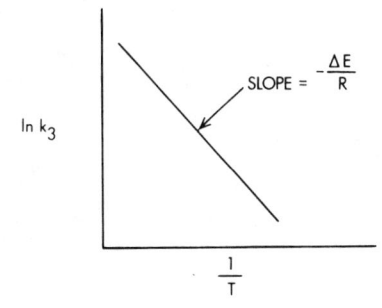

Figure 2.8. The Arrhenius plot; slope equals $-\Delta E/R$.

perimental activation energy can be related to $\Delta H^\ddagger$ and $\Delta S^\ddagger$, which are the enthalpy and entropy changes associated with the activation of ES to ES‡, the transition state. Since $\Delta G^\ddagger = \Delta H^\ddagger - T\Delta S^\ddagger$, Eq. 9 can be transformed as follows:

$$k_3 = \frac{kT}{h} e^{-\Delta G^\ddagger/RT} \quad (9)$$

$$k_3 = \frac{kT}{h} e^{\Delta S^\ddagger/R} e^{-\Delta H^\ddagger/RT} \quad (53)$$

Differentiation of Eq. 52a with respect to T gives

$$d \ln k_3/dT = -\Delta E/RT^2 \quad (54)$$

Since $\Delta G^{\ddagger} = -RT \ln K^{\ddagger}$, Eq. 9 can be written as

$$k_3 = \frac{kT}{h} K^{\ddagger} \qquad (55)$$

Taking logs and differentiating with respect to T gives

$$\frac{d \ln k_3}{dT} = \frac{1}{T} + \frac{d \ln K^{\ddagger}}{dT} \qquad (56)$$

Substitution of the van't Hoff equation, $d \ln K^{\ddagger}/dT = \Delta H^{\ddagger}/RT^2$, gives

$$\frac{d \ln k_3}{dT} = \frac{1}{T} + \frac{\Delta H^{\ddagger}}{RT^2} \qquad (57)$$

Combination of Eq. 57 with Eq. 54 gives

$$\frac{\Delta E}{RT^2} = \frac{1}{T} + \frac{\Delta H^{\ddagger}}{RT^2} \qquad (58)$$

or

$$\Delta E = RT + \Delta H^{\ddagger}$$

Hence, the enthalpy of activation can be calculated from the experimental activation energy. The value of $\Delta S^{\ddagger}$ can then be calculated, using Eq. 53.

The values of $\Delta H^{\ddagger}$ and $\Delta S^{\ddagger}$ can be important in mechanistic interpretation. Consider the comparison of the activities of a chemically modified and a native enzyme. Use of the activation parameters can provide insight into whether the modification results in an entropic effect (such as orientation) or an energetic effect (such as strain). Deductions based solely on changes in activation parameters are not conclusive, but can be useful when combined with other findings.

The final effect of temperature on enzyme-catalyzed reactions is thermal denaturation of enzyme, which ordinarily occurs in the range of 55-65°C. This range is certainly not applicable to all enzymes: some denature at lower temperatures. Enzymes from thermophilic bacteria, which live in hot springs, show an improved tolerance to heat, some remaining active after exposure to temperatures over 90°C. This thermal stability is of great academic and practical interest. What structural features are responsible for it? Perutz has described the comparison of the structure of normal glyceraldehyde phosphate dehydrogenase with that of the enzyme from the thermophile *Bacillus stereothermophilus* (10). Structural studies of the normal and the *Bacillus* enzyme revealed that salt bridges between subunits occurred only in the thermostable enzyme (11-12). These salt bridges are shielded from water and so are of sufficient strength to provide the additional thermal stability of the *Bacillus* enzyme.

E. INHIBITION

Noncovalent

There are three basic types of noncovalent inhibition—competitive, uncompetitive, and noncompetitive. Inhibitors in these categories can be substrates, products, or other molecules. The distinctions among the types of inhibition rest on the enzyme form with which the inhibitor combines and whether binding of inhibitor prevents binding of substrate. Diagnosis of the type of inhibition is accomplished with the Lineweaver-Burk plot. Competitive inhibition produces only a slope effect; for uncompetitive inhibition there is only an intercept effect. Noncompetitive inhibition produces both.

In the simplest case of *competitive inhibition* the substrate and inhibitor compete for the same binding site. But there are other models that embody the feature of mutually exclusive binding of inhibitor and substrate (Fig. 2.9). An excess of substrate reverses the inhibition, which means V_m is unchanged. Consider the following scheme:

$$E + S \underset{K_m}{\rightleftharpoons} ES \xrightarrow{k_{cat}} E + P$$

$$E + I \underset{K_1}{\rightleftharpoons} EI; \quad K_1 = \frac{[E] I_0}{[EI]} \tag{59}$$

Since $[EI] \ll [I]$, $[I]$ is represented by I_0, the total amount of inhibitor in the system. The rate equation for competitive inhibition is

$$v_0 = \frac{V_m S_0}{S_0 + K_m + K_m I_0/K_i} = \frac{V_m S_0}{S_0 + K_m(1 + I_0/K_i)} \tag{60}$$

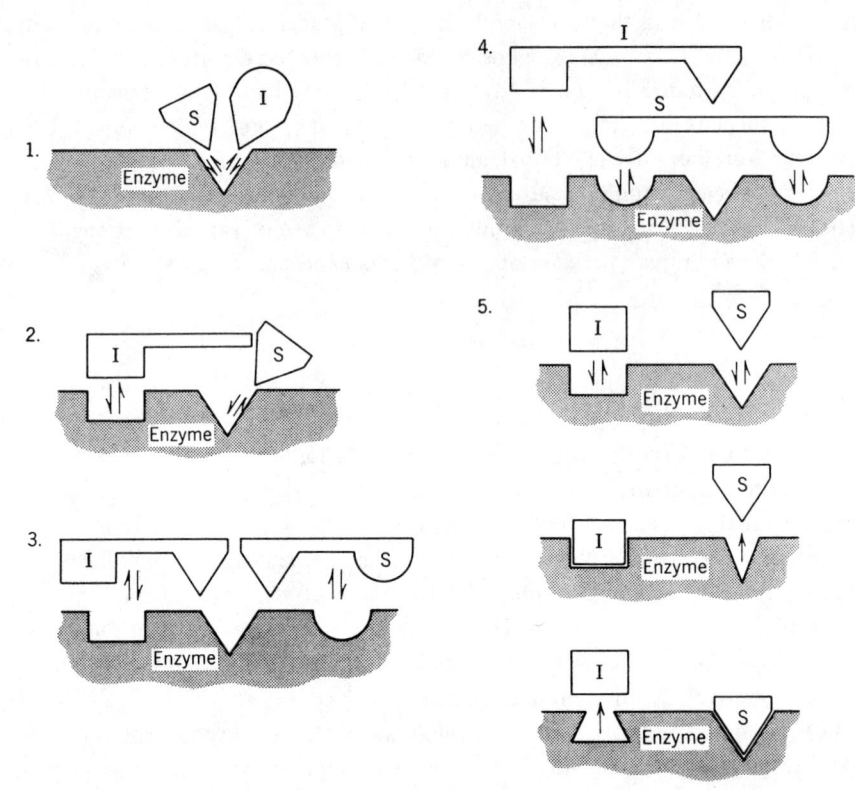

Figure 2.9. Models for the action of a competitive inhibitor. In all cases S and I bind in a mutually exclusive fashion. From Segel, *Enzyme Kinetics*, Wiley, New York (1975), p. 102. Reproduced by permission.

K_m is increased by the factor $(1 + I_0/K_i)$. The degree of inhibition is dependent on the relative binding affinities of substrate and inhibitor (K_m/K_I), along with the relative amounts of substrate and inhibitor present. The double reciprocal form of Eq. 60 is

$$\frac{1}{v_0} = \frac{K_m(1 + I_0/K_I)}{V_m} \frac{1}{S_0} + \frac{1}{V_m} \tag{61}$$

Plots of $1/v_0$ versus $1/S_0$ at various values of I_0 result in a family of straight lines, intersecting on the $1/v_0$ axis (Fig. 2.10). The slopes of the lines equal $K_m(1 + I_0/K_I)/V_m$ or

E. INHIBITION

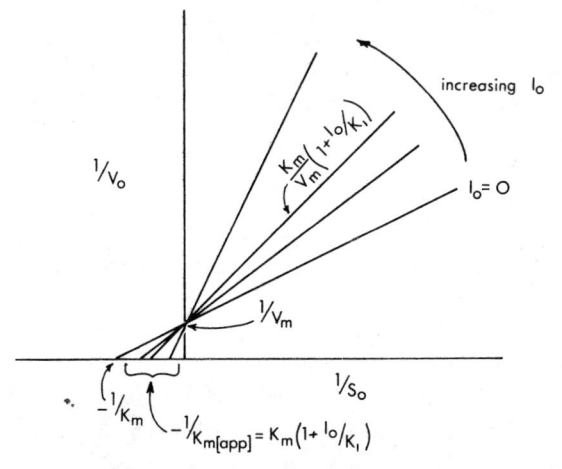

Figure 2.10. Double reciprocal plots at varying $[I_0]$, where I is a competitive inhibitor.

$$\text{slope} = \frac{K_m}{V_m} + \frac{K_m}{V_m K_I} \cdot (I_0) \tag{62}$$

A slope replot based on Eq. 62 will give a straight line having an I_0 intercept equal to $-K_I$ (Fig. 2.11). The intercept on the vertical axis is equal to K_m/V_m. This value should be equal to that obtained for the slope of the double reciprocal plot in $I_0 = 0$.

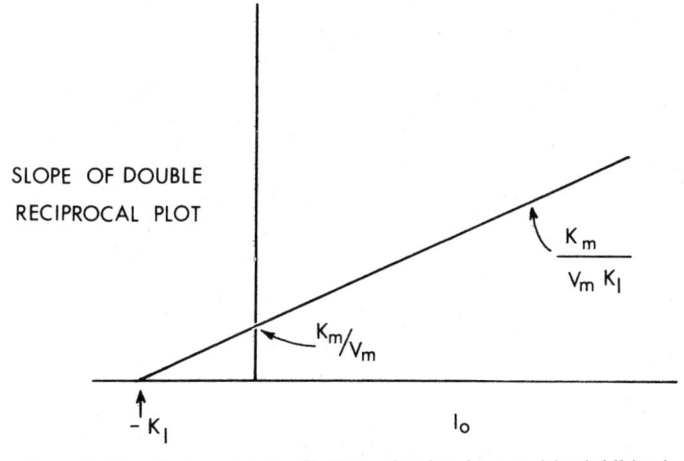

Figure 2.11. Slope replot for K_I determination (competitive inhibitor).

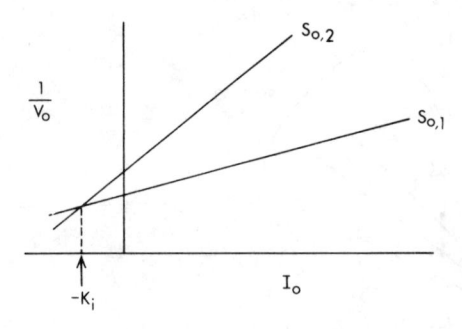

Figure 2.12. Dixon plot for K_I determination.

An alternative approach to the evaluation of K_I is based on the Dixon plot, $1/v_0$ versus I_0. In this case substrate concentration in a given experiment is held constant and the inhibitor concentration is varied. The dependence of v_0 on I_0 must be found for at least two substrate concentrations. The point of intersection of two plots of $1/v_0$ versus I_0 at two substrate concentrations is equal to $-K_I$ (Fig. 2.12). The proof of this relationships follows. At the point of intersection

$$1/v_{0,1} = 1/v_{0,2} \tag{63}$$

Substitution of Eq. 61 gives

$$\frac{K_m}{V_m}(1 + I_0/K_I)\frac{1}{S_{0,1}} + \frac{1}{V_m} = \frac{K_m}{V_m}(1 + I_0/K_i)\frac{1}{S_{0,2}} + \frac{1}{V_m} \tag{64}$$

Subtraction of $1/V_m$ from both sides and multiplying through by V_m/K_m yields

$$(1 + I_0/K_I)\frac{1}{S_{0,1}} = (1 + I_0/K_I)\frac{1}{S_{0,2}} \tag{65}$$

The quantity $(1 + I_0/K_I)$ is a factor on both sides. However, $1/S_{0,1}$ does not equal $1/S_{0,2}$, which means that for the equality to hold, $(I + I_0/K_I)$ must equal zero. Therefore, at the point of intersection

$$I + I_0/K_I = 0; I_0 = -K_I \tag{66}$$

E. INHIBITION

Uncompetitive inhibition occurs when the inhibitor binds exclusively to the ES complex with the formation of an inactive ESI complex:

$$E + S \underset{K_m}{\rightleftarrows} ES \rightarrow E + P$$

$$ES + I \underset{K_I}{\rightleftarrows} ESI$$

The reciprocal form of the rate equation is

$$1/v_0 = \frac{K_m}{V_m} 1/S_0 + \frac{(1 + I_0/K_I)}{V_m} \qquad (67)$$

Only the intercept is a function of I_0, which means parallel lines will result when $1/v_0$ is plotted versus $1/S_0$ at varying values of I_0 (Fig. 2.13). The intercepts on the horizontal axis equal $(1 + I_0/K_I)/K_m$ or

$$\frac{1}{K_{m(app)}} = \frac{I_0}{K_I K_m} + \frac{1}{K_m} \qquad (68)$$

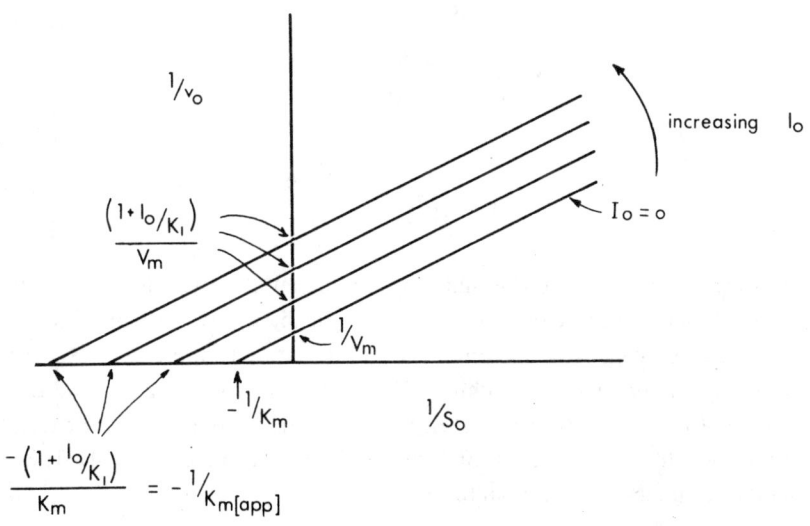

Figure 2.13. Double reciprocal plots of varying concentrations of an *uncompetitive* inhibitor.

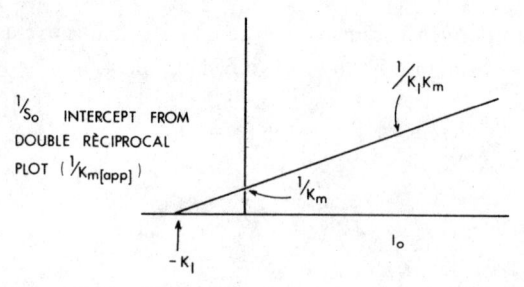

Figure 2.14. Slope replot for *uncompetitive* inhibition.

The replot for K_I determination is $1/K_{m(app)}$ versus I_0 (Fig. 2.14).

Noncompetitive inhibition can be represented by the scheme below

$$E + S \underset{K_m}{\rightleftarrows} ES \xrightarrow{k_{cat}} E + P$$

$$E + I \underset{K_I}{\rightleftarrows} EI$$

$$ES + I \underset{K_I}{\rightleftarrows} ESI \quad \text{or}$$

$$\begin{array}{c} E + S \underset{K_m}{\rightleftarrows} ES \xrightarrow{k_{cat}} E + P \\ + \quad\quad\quad + \\ I \quad\quad\quad I \\ \updownarrow K_I \quad\quad K_I \updownarrow \\ EI + S \underset{K_m}{\rightleftarrows} ESI \end{array}$$

$$K_I = \frac{[E]\,I_0}{[EI]} = \frac{[ES]\,I_0}{[ESI]}$$

In this case E and ES bind the inhibitor with equal affinity, and E and EI bind substrate equally well. This condition results in the intersection of $1/v_0$ versus $1/s_0$ plots at varying I_0, as shown in Figure 2.15. Stated another way, K_m is not changed by the presence of inhibitor. V_m is changed, since part of E and ES are tied up as EI and ESI. It can be seen in the scheme above that a great excess of substrate would not convert all of E into the ES form (ES can bind I). The rate equation for noncompetitive inhibition is

$$1/v_0 = \frac{K_m\,(1 + I_0/K_I)}{V_m} \cdot \frac{1}{S_0} + \frac{(1 + I/K_I)}{V_m} \tag{68}$$

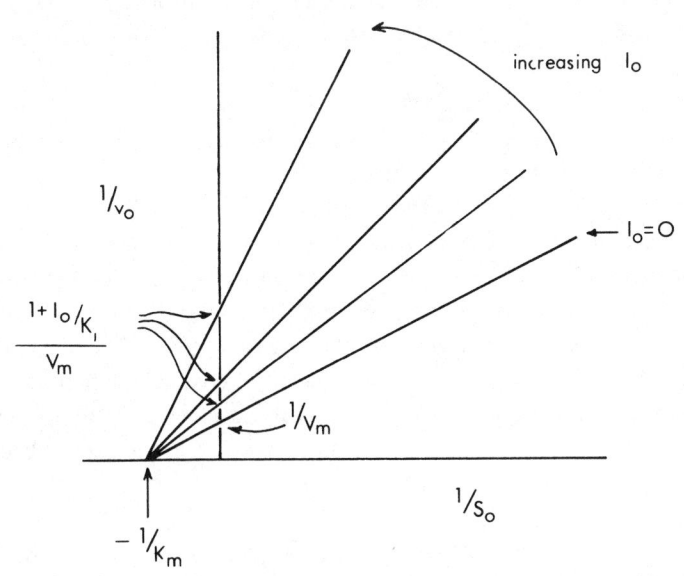

Figure 2.15. Double reciprocal plots for intersecting noncompetitive inhibition.

K_I can be determined from a replot of the slopes of the double reciprocal plots, as described for competitive inhibition (Fig. 2.11).

As stated above, noncompetitive inhibition results in effects on both the slope and the intercept. In the classic case of "intersecting" noncompetitive inhibition, discussed above, the lines intersect on the horizontal axis. In general the lines may cross above or below the horizontal axis. In some cases none of the lines intersects at the same point. The term "mixed" inhibition was formerly used for those cases where intersection does not occur on the horizontal axis, but Cleland points out that there is no good theoretical reason for this distinction (14).

Covalent Inhibition

Many enzymes are inhibited by heavy metals. This type of inhibition involves formation of metal sulfides with essential thiol groups:

$$E\text{-}SH + HgCl_2 \rightarrow E\text{-}S\text{-}Hg\text{-}Cl$$

Alkylating agents, acylating reagents, strong oxidants, and reducing agents can bring about covalent modification of the enzyme with loss of activity. As we shall see in the next chapter, results with covalent inhibitors are often valuable in mechanistic studies.

The effect of a covalent, irreversible inhibitor is a reduction of the amount of active enzyme present, which results in a reduction of $V_{m(app)}$. The double reciprocal plots for classical noncompetitive and irreversible inhibition are identical in form. This similarity has led to the classification of irreversible, covalent inhibitors as noncompetitive. The designation is incorrect because of the covalent nature of the bond between enzyme and inhibitor. The inhibition is irreversible in the sense that conventional methods of separating enzymes from small molecules (e.g., dialysis) do not regenerate the original activity. The distinction between noncompetitive inhibition and covalent, irreversible inhibition can be made easily with the aid of a plot of $V_{m(app)}$ versus E_0 (Fig. 2.16). For the non-covalent inhibitor, the slope of the plot will decrease from the control because

$$V_{m(app)} = \frac{k_{cat} E_0}{(1 + I_0/K_I)} \qquad (69)$$

This plot will go through the origin. In the case of an irreversible inhibitor, the plot will have a finite intercept on the E_0 axis that corresponds to the amount of enzyme reacting with the covalent inhibitor.

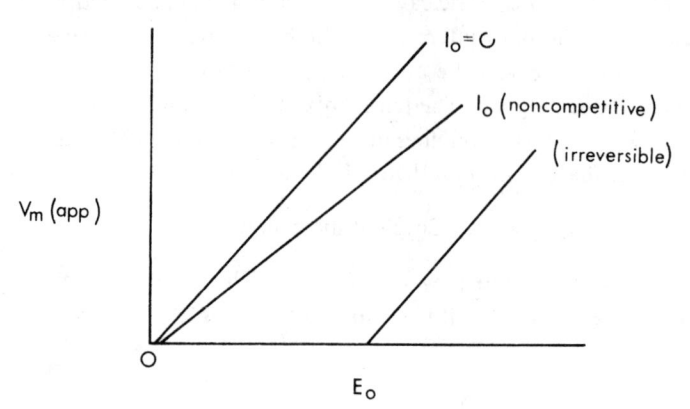

Figure 2.16. Plot of V_m versus E_0 to detect irreversible inhibition, such as poisoning by sulfide formation.

F. BIREACTANT SYSTEMS

Many enzyme-catalyzed reactions involve more than one substrate. Until the 1950s most enzymologists studied the kinetics of hydrolytic enzymes because they were readily available. These reactions do involve two substrates, but one substrate, water, is present in such great excess that the kinetic equations reduce to Henri-Michaelis-Menten form. Steady-state equations were derived by Alberty, Dalziel, Wong, Hanes, and others for multisubstrate reactions, including

$$A + B \rightleftharpoons P + Q \tag{70}$$

In 1963 Cleland introduced a clear, general procedure for deriving rate equations for multisubstrate reactions and a shorthand method for representing the various pathways. In this section we shall consider bireactant systems only. Four common pathways will be examined in detail—ping-pong, ordered, Theorell-Chance, and random-rapid equilibrium. These pathways are represented below with the conventional and Cleland shorthand schemes:

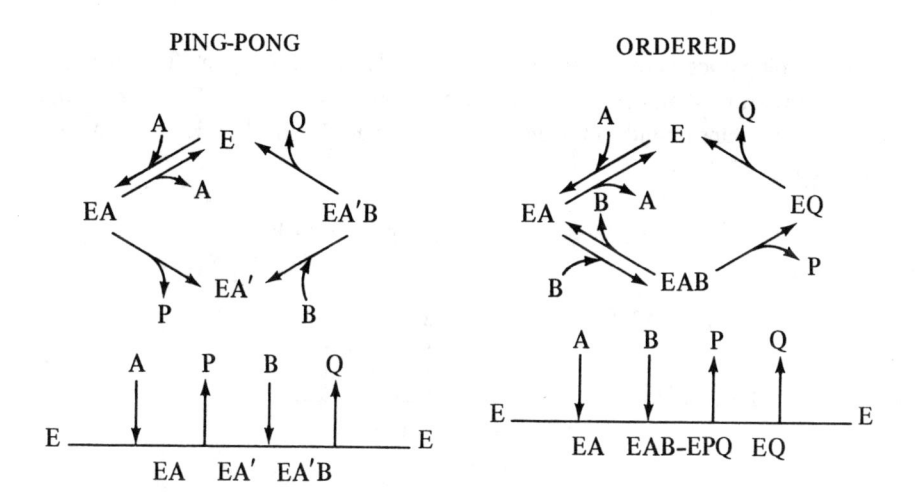

KINETICS OF ENZYME-CATALYZED REACTIONS

THEORELL-CHANCE

```
      A   E   Q
       ↘ ↗ ↙
         A
EA ─────────→ EQ
     B     P
```

```
    A B     P     Q
    ↓ ↓     ↑     ↑
E ──────────────────── E
    EA      EQ
```

RANDOM-RAPID EQUILIBRIUM

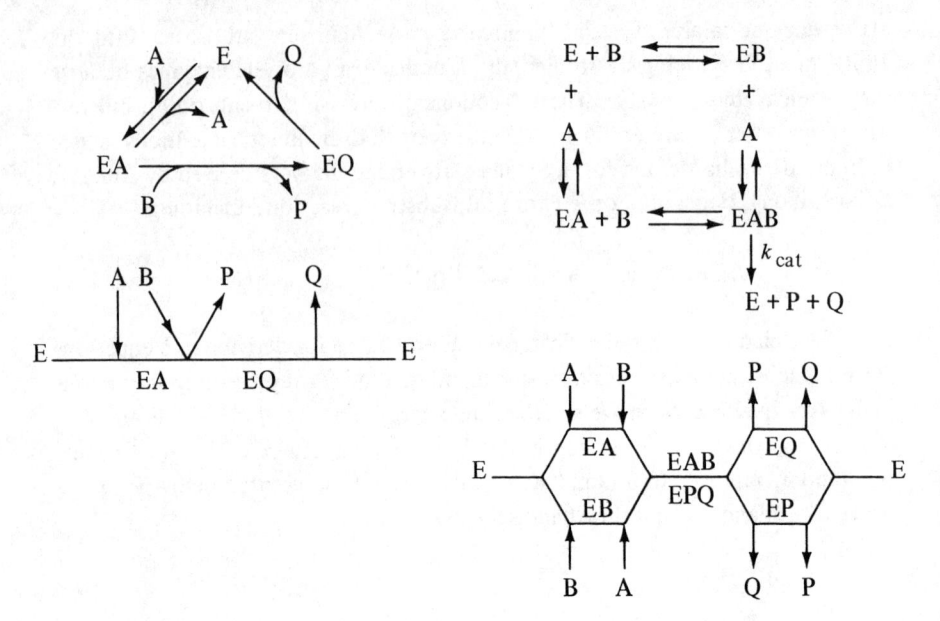

$$E + B \rightleftharpoons EB$$
$$+ \qquad \qquad +$$
$$A \updownarrow \qquad A \updownarrow$$
$$EA + B \rightleftharpoons EAB$$
$$\downarrow k_{cat}$$
$$E + P + Q$$

```
       A B           P   Q
        ↓↓           ↑   ↑
       ╱EA╲   EAB   ╱EQ╲
E ────         ────        ──── E
       ╲EB╱   EPQ   ╲EP╱
        ↑↑           ↓   ↓
       B A           Q   P
```

In the ping-pong pathway there are two displacement reactions. The first step is the formation of an EA complex, which breaks down to P and EA′. Part of the substrate remains bound to the enzyme; the remaining part, P, is released. A general example follows.

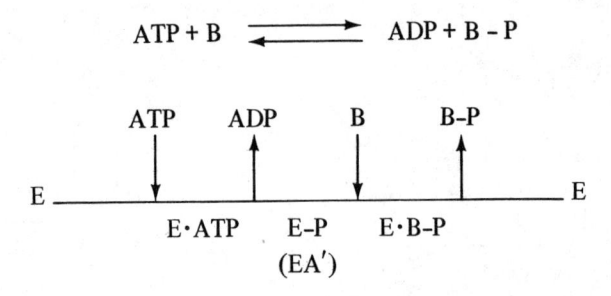

$$ATP + B \rightleftharpoons ADP + B-P$$

```
     ATP    ADP    B     B-P
      ↓      ↑     ↓      ↑
E ──────────────────────────── E
     E·ATP   E-P    E·B-P
             (EA′)
```

In the first chemical step, a phosphate group of ATP is transferred to the enzyme with the formation of ADP. The phosphoryl enzyme intermediate then transfers the phosphate group to the acceptor, B.

F. BIREACTANT SYSTEMS

The steady-state rate equation can be derived by the King-Altman method. The steps involving products are considered irreversible since products are present in very low amounts in initial velocity studies. The $n-1$ lined patterns are:

E

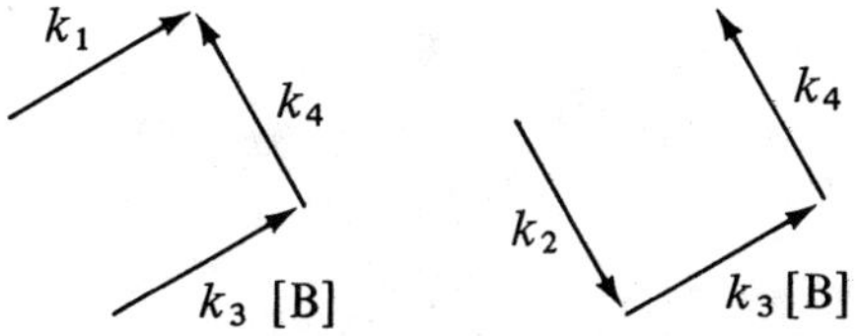

k_1, k_4, $k_3[B]$, k_2, k_4, $k_3[B]$

EA

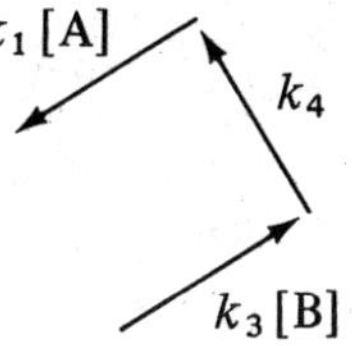

$k_1[A]$, k_4, $k_3[B]$

EA'

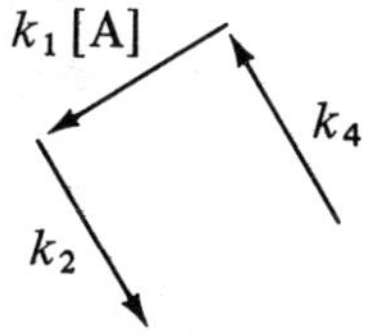

$k_1[A]$, k_4, k_2

EA'B

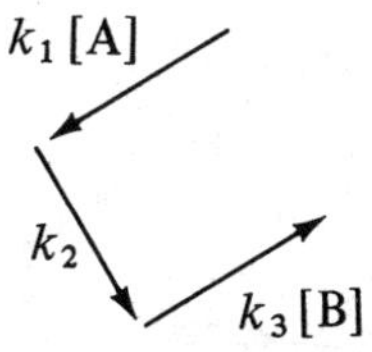

$k_1[A]$, k_2, $k_3[B]$

The distribution equations are therefore

$$\frac{E}{E_0} = \frac{k_1 k_3 k_4 [B] + k_2 k_3 k_4 [B]}{\Sigma \text{ all terms}} \tag{71}$$

$$\frac{[EA]}{E_0} = \frac{k_1 k_3 k_4 [A][B]}{\Sigma \text{ all terms}} \tag{72}$$

$$\frac{[EA']}{E_0} = \frac{k_1 k_2 k_4 [A]}{\Sigma \text{ all terms}} \tag{73}$$

$$\frac{[EA'B]}{E_0} = \frac{k_1 k_3 k_4 [A][B]}{\Sigma \text{ all terms}} \tag{74}$$

$$\text{rate} = k_4 [EA'B] \tag{75}$$

From the distribution equations it can be shown that

$$[EA'B] = \frac{k_1 k_2 k_3 E_0 [A][B]}{k_1 k_2 k_4 [A] + k_3 k_4 (k_{-1} + k_2)[B] + k_1 k_3 (k_2 + k_4)[A][B]} \tag{76}$$

Combination of Eqs. 75 and 76 gives

$$v = \frac{k_2 k_4 E_0 [A][B]/(k_2 + k_4)}{k_2 k_4 [A]/k_3 (k_2 + k_4) + k_4 (k_{-1} + k_2)/k_1 (k_2 + k_4)[B] + [A][B]} \tag{77}$$

or

$$v_0 = \frac{V_m A_0 B_0}{K_{m,B} A_0 + K_{m,A} B_0 + A_0 B_0} \tag{78}$$

The reciprocal form of Eq. 78 is

$$\frac{1}{v_0} = \frac{K_{m,A}}{V_m} \frac{1}{A_0} + \frac{1}{V_m} \frac{1 + K_m B}{B_0} \tag{79}$$

Plots of $1/v_0$ versus $1/A_0$ at various fixed levels of B_0 will yield a family of parallel lines, as is shown is Figure 2.17a. The slopes of the lines are equal to $K_{m,A}/$

F. BIREACTANT SYSTEMS

V_m. The intercepts on both axes are functions of B_0: $1/V_{m(\text{app})} = (1 + K_{m,B}/B_0)/V_m$ and $-1/K_{m(\text{app})} = -(1 + K_{m,B}/B_0)/K_{m,B}$. When the fixed level of B_0 is sufficiently high that $K_{m,B}/B_0 \ll 1$, the limiting values of the intercepts are $1/V_m$ and $-1/K_{m,B}$, respectively. One approach to the evaluation of the constants in Eq. 79 entails the use of the intercept replot, $1/V_{m(\text{app})}$ versus $1/B_0$, shown in Figure 2.17b. The reciprocal of V_m is the intercept on the vertical axis

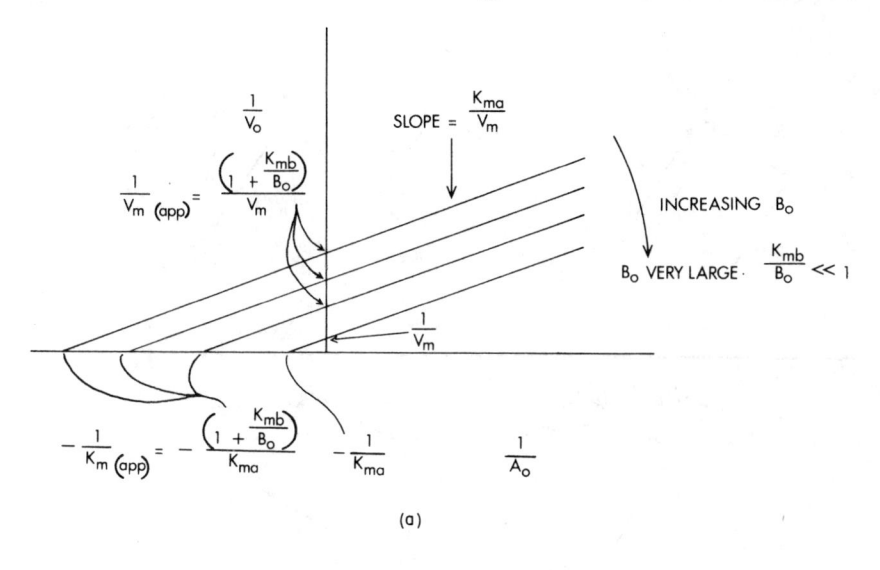

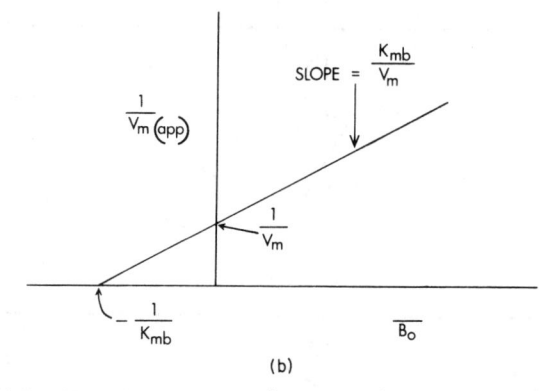

Figure 2.17. (a) Double reciprocal plots ($1/v_0$ versus $1/A_0$) at various concentrations of B. The parallel lines are consistent with the ping-pong pathway. (b) Intercept replot, using intercepts from plots similar to those found in (a).

of the replot. Since the slopes of the original double reciprocal plots equal $K_{m,A}/V_m$, $K_{m,A}$ can be calculated. The $1/B_0$ intercept of the replot yields $K_{m,B}$.

In the ordered mechanism shown above, the ternary complex EAB is formed only from the binary complex EA. The enzyme intermediates are E, EA, EAB, and EQ. The $n-1$ lined patterns for these intermediates follow.

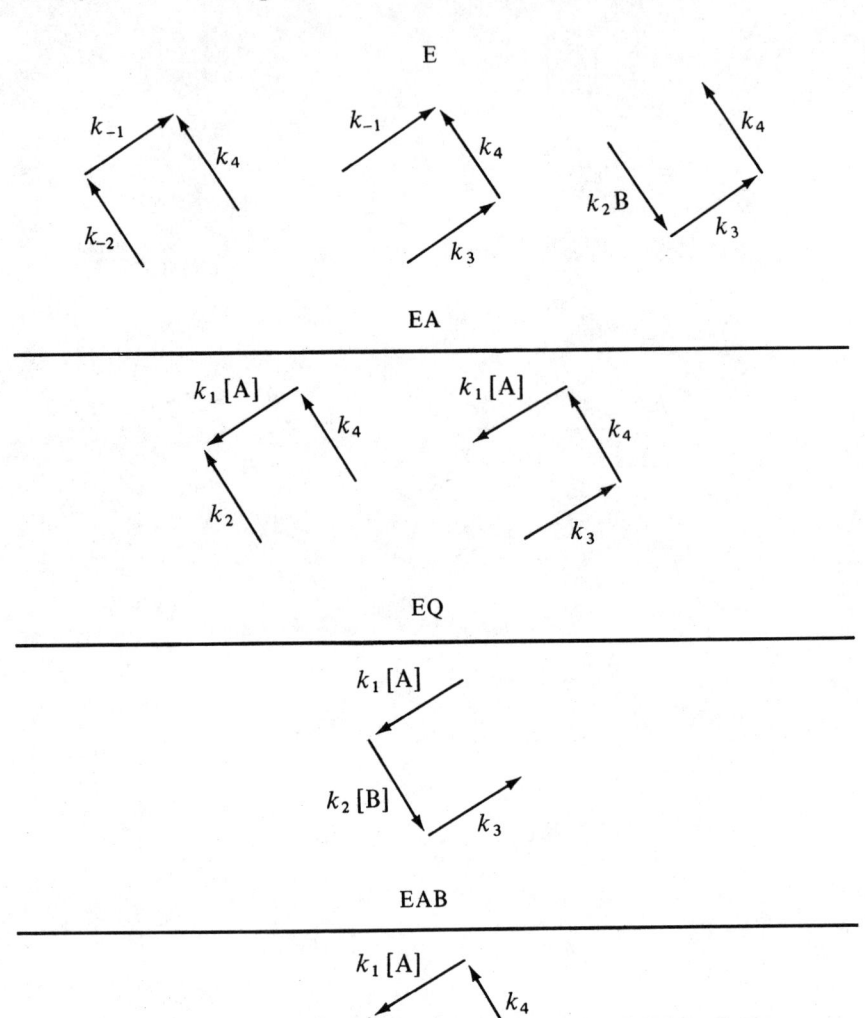

E

EA

EQ

EAB

F. BIREACTANT SYSTEMS

The distribution equations are therefore

$$\frac{E}{E_0} = \frac{k_{-1}k_{-2}k_4 + k_{-1}k_3k_4 + k_2k_3k_4 [B]}{\Sigma \text{ all terms}} \quad (80)$$

$$\frac{[EA]}{E_0} = \frac{k_1k_{-2}k_4 [A] + k_1k_3k_4 [A]}{\Sigma \text{ all terms}} \quad (81)$$

$$\frac{[EQ]}{E_0} = \frac{k_1k_2k_3 [A][B]}{\Sigma \text{ all terms}} \quad (82)$$

$$\frac{[EAB]}{E_0} = \frac{k_1k_2k_4 [A][B]}{\Sigma \text{ all terms}} \quad (83)$$

$$\text{rate} = k_3 [EAB] \quad (84)$$

Using Eqs. 80-83, it can be shown that

$$EAB = \frac{k_1k_2k_4 E_0 [A][B]}{k_4(k_{-1}k_{-2} + k_{-1}k_3 + k_2k_3 [B] + k_1k_4(k_{-2} + k_3)[A] + k_1k_2(k_3 + k_4)[A][B]} \quad (85)$$

The initial velocity is given by

$$v_0 = \frac{[k_3k_4/(k_3 + k_4)] E_0 A_0 B_0}{k_{-1}k_4(k_{-2} + k_3)/k_1k_2(k_3 + k_4) + k_4(k_{-2} + k_3)A_0/k_2(k_3 + k_4) + k_3k_4 B_0/k_1(k_3 + k_4) + A_0 B_0} \quad (86)$$

or

$$v_0 = \frac{V_m A_0 B_0}{K_{i,a} K_{m,B} + K_{m,B} A_0 + K_{m,A} B_0 + A_0 B_0} \quad (87)$$

in which $K_{i,a} = k_{-1}/k_1$. Comparison of Eq. 87 with Eq. 78 reveals that the only difference in form is the $K_{i,a} K_{m,B}$ term in the denominator. The double reciprocal form of Eq. 87 is

$$\frac{1}{v_0} = \frac{K_{m,A}}{V_m} \left(1 + \frac{K_{i,a}K_{m,B}}{K_{m,A}B_0}\right) \frac{1}{A_0} + \frac{1}{V_m} \left(1 + \frac{1 + K_{m,B}}{B_0}\right) \quad (88)$$

KINETICS OF ENZYME-CATALYZED REACTIONS

Plots of $1/v_0$ versus $1/A_0$ at various fixed levels of B_0 will intersect to the left of the vertical axis (Fig. 2.18a). The $1/v_0$ coordinate of the point of intersection is given by $1/V_m(1 - K_{m,a}/K_{i,a})$, which means that the lines may intersect above or below the $1/A_0$ axis. The $1/A_0$ coordinate of the point of intersection is equal to $-K_{i,a}$. A replot of $1/v_{m(app)}$ versus $1/B_0$ appears in Figure 2.18b. V_m and

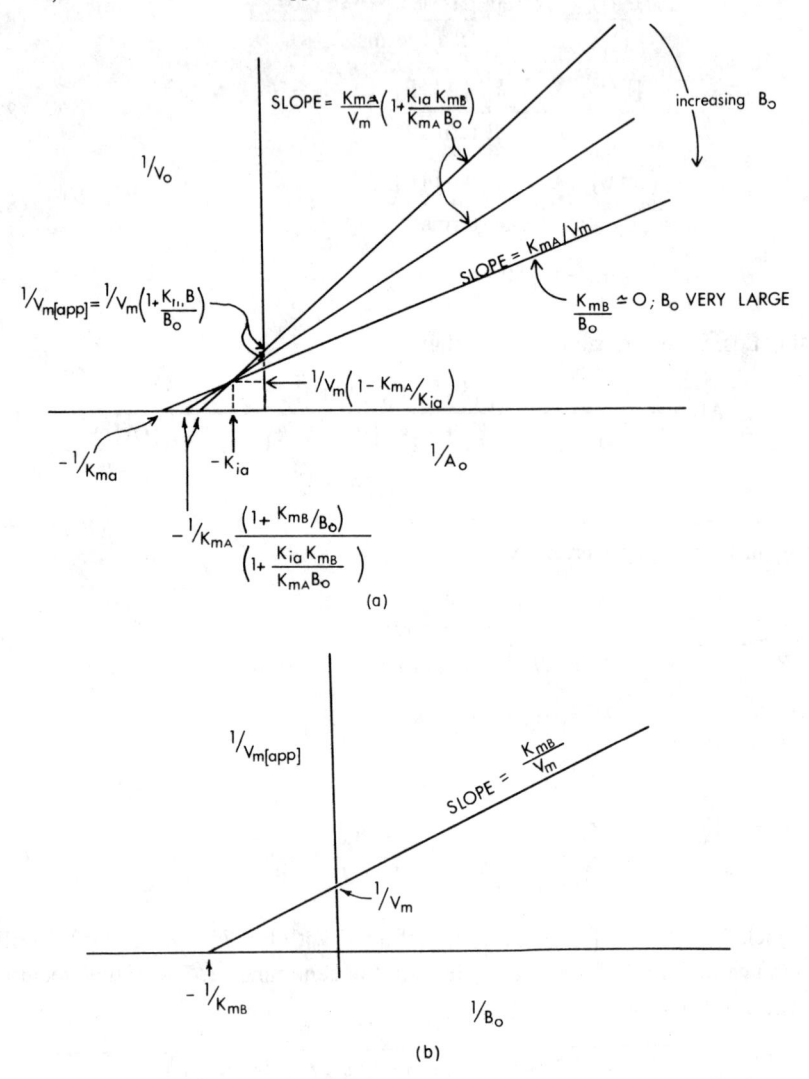

Figure 2.18. (a) Double reciprocal plots for ordered, random, rapid-equilibrium, and Theorell-Chance mechanisms. (b) Intercept replot for above cases.

F. BIREACTANT SYSTEMS

$K_{m,B}$ can be calculated by using the intercepts of the replot. $K_{i,a}$ is gotten from the $1/A_0$ coordinate of the point of intersection in Figure 2.18a. The remaining constant, $K_{m,a}$, can then be calculated from the slopes or $1/A_0$ intercepts of the original reciprocal plot (Fig. 2.18a).

The Theorell-Chance mechanism is a special form of the ordered mechanism. A adds first to form the EA complex, as before. EA then reacts with B to yield EQ + P without the formation of a ternary complex of sufficient lifetime to be kinetically significant.

The King's patterns are

E

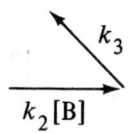

k_{-1} k_3 k_3

$k_2[B]$

EA

$k_1[A]$ k_3

EQ

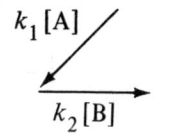

$k_1[A]$

$k_2[B]$

The distribution equations are

$$\frac{[E]}{E_0} = \frac{k_{-1}k_3 + k_2k_3\,[B]}{\Sigma \text{ all terms}} \tag{89}$$

$$\frac{[EA]}{E_0} = \frac{k_1 k_3 [A]}{\Sigma \text{ all terms}} \tag{90}$$

$$\frac{[EQ]}{E_0} = \frac{k_1 k_2 [A]\,[B]}{\Sigma \text{ all terms}} \tag{91}$$

The rate is given by

$$\text{rate} = k_3 [EQ] \tag{92}$$

Hence

$$v_0 = \frac{k_1 k_2 k_3 \, E_0 A_0 B_0}{k_{-1}k_3 + k_1 k_3 A_0 + k_2 k_3 B_0 + k_1 k_2 \, A_0 B_0} \tag{93}$$

Division of top and bottom by $k_1 k_2$ yields

$$v_0 = \frac{k_3 E_0 A_0 B_0}{k_{-1}/k_1 \, k_3/k_2 + k_3/k_2 \, A_0 + k_3/k_1 \, B_0 + A_0 B_0} \tag{94}$$

or

$$v_0 = \frac{V_m \, A_0 \, B_0}{K_{i,a} K_{m,B} + K_{m,B} A_0 + K_{m,A} B_0 + A_0 B_0} \tag{95}$$

which is identical in form to Eq. 87. In Eq. 95 $K_{m,B} = k_3/k_2$, $K_{m,A} = k_3/k_1$, and $V_m = k_3 E_0$. These terms follow directly from Eq. 87 when $k_3 \gg k_{-2}$ and $k_4 \gg k_3$. The reciprocal plot and replot for the Theorell-Chance system are identical to those for the ordered pathway.

In the random-rapid equilibrium pathway, k_{cat} is small, so that EA, EB, and EAB are in equilibrium. This assumption simplifies the derivation of the rate equation. The equilibrium expressions are

$$K_A' = \frac{[E][A]}{[EA]} \; ; K_B' = \frac{[E][B]}{[EB]} \; ; K_A = \frac{[EB][A]}{[EAB]} \; ; K_B = \frac{[EA][B]}{[EAB]} \tag{96}$$

The enzyme conservation law is

$$E_0 = [E] + [EA] + [EB] + [EAB] \tag{97}$$

The rate is given by

$$\text{rate} = k_{\text{cat}} [EAB] \tag{98}$$

F. BIREACTANT SYSTEMS

Solving Eqs. 96 and 97 for [EAB] gives

$$[EAB] = \frac{[A][B]E_0/K_A'K_B}{1 + [A]/K_A' + [B]/K_B + [A][B]/K_A'K_B} \tag{99}$$

The rate equation is therefore

$$v_0 = \frac{V_m A_0 B_0}{K_A'K_B + K_B A_0 + K_A' B_0 + A_0 B_0} \tag{100}$$

Eq. 100 is identical in form to Eqs. 95 (Theorell-Chance) and 87 (ordered). The reciprocal plot and replot are similar to those described for the ordered pathway (Eq. 88, Fig. 2.18).

It is obvious that initial velocity dependence on substrate concentrations cannot be used to delineate among the four mechanisms given above. Even the ping-pong mechanism can be confused with the ordered mechanism. When $K_{i,a}$ in Eq. 88 is large in comparison to $K_{m,A}$, the sensitivity of the slopes of the reciprocal plots to the concentration of fixed substrate, B_0, can be sufficiently small to produce nearly parallel lines. Therefore, the observation of parallel reciprocal plots cannot be used as the sole proof of the ping-pong mechanism. Clearly, information other than reciprocal plots is needed to distinguish among ordered, Theorell-Chance, and random mechanisms.

Inhibition and isotope exchange studies are the other tools used by kineticists in conjunction with the dependence of initial rates on substrate concentration. Product inhibitor studies are very useful. The product inhibitor works by "backing up" the reaction. If an EPQ complex normally dissociates to P and EQ, the presence of added P in the system will tend to reverse this step. The "dead-end" inhibitor is not normally associated with products of the reaction, but it is capable of tying up one or more enzyme forms. A "dead-end" inhibitor, I, could bind to EQ to form EQI, which cannot react further except to dissociate into EQ and I. The presence of I slows the reaction, but not by backing it up, as in the case of a product inhibitor. As was mentioned earlier, replots can be made to determine the magnitude of the various inhibition constants. However, in some cases with multisubstrate reactions these replots are not linear, but parabolic or hyperbolic. Parabolic inhibition is caused by a product's binding to more than one enzyme form; hyperbolic replots result when alternate reaction sequences occur. We shall not consider these cases involving curved replots, but the reader is referred to reference 16 for a more detailed discussion.

Characteristic effects of double reciprocal plots on slopes and intercepts are brought about by product inhibitors. We shall use Cleland's rules to interpret product inhibition patterns.

Rule 1. The *intercept* of a reciprocal plot is changed by an inhibitor that associates with an enzyme form *other than* the one with which the variable substrate combines (*uncompetitive inhibition*).

Rule 2. A *slope* effect occurs when the inhibitor combines with an enzyme form that is the same as, or is connected by a series of reversible steps to, the enzyme form with which the variable substrate combines (*competitive inhibition*).

Notes:

1. Noncompetitive inhibition results when both rules apply.

2. An inhibitor cannot bind to an enzyme form having a steady state level that is already zero because of saturation by a substrate.

3. No intercept effects occur when rapid-equilibrium conditions prevail.

The procedure for application of these rules is straightforward, as we shall see. The first step is to write down the enzyme forms to which the inhibitor and variable substrate bind. If the enzyme forms are the same, a slope effect results (competitive inhibition). If the forms are different, check for reversible connections; if these are found, the inhibitor is noncompetitive. If no reversible connections are found, the inhibition is uncompetitive. The presence of the fixed substrate at saturating levels produces two possible consequences. First, the level of the enzyme form to which the inhibitor normally binds can be reduced to zero, resulting in no inhibition. Reversible connections can also be broken by reduction of the level of an enzyme form within the connection.

Let us now apply Cleland's rules to predict product inhibition patterns for the ordered mechanism $(A + B \rightleftharpoons P + Q)$.

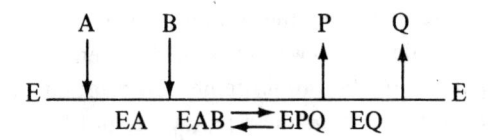

$$\begin{array}{c} A \quad B \quad\quad P \quad Q \\ \downarrow \quad \downarrow \quad\quad \uparrow \quad \uparrow \\ E \underline{\quad\quad\quad\quad\quad\quad\quad\quad\quad\quad} E \\ \quad EA \quad EAB \rightleftharpoons EPQ \quad EQ \end{array}$$

Referring to Table 2.1, consider inhibition by P, with A as the variable substrate; B is present at a level below saturation. P will tend to "back up" the reaction by binding to EQ. A binds to E. Since EQ and E are different, an intercept effect

TABLE 2.1. Product Inhibition Patterns for the Ordered Pathway (A + B ⇌ P + Q)

```
         A      B          P     Q
         ↓      ↓          ↑     ↑
    E ──────────────────────────────── E
         EA   EAB ⇌ EPQ   EQ
```

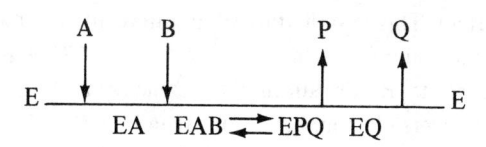

Inhibitor	Variable Substrate	Saturating Substrate	Type of Inhibition
P	A	–	P binds to EQ, A to E: intercept effect. Reversible connection produces slope effect (noncompetitive)
P	A	B	P binds to EQ, A to E: intercept effect. Reduction of EA to zero breaks reversible connection (uncompetitive)
Q	A	–	Q and A both bind to E: slope effect (competitive)
Q	A	B	Same as above (competitive)
P	B	–	P binds to EQ, B to EA: intercept effect. Reversible connection gives slope effect (Noncompetitive)
P	B	A	Same as above (noncompetitive)
Q	B	–	Q binds to E, B to EA: intercept effect. Reversible connection gives slope effect (noncompetitive)
Q	B	A	Q binds to E, which is reduced to zero by saturating A (no inhibition)

occurs. EQ and E are connected reversibly because there are no intervening product release steps or very low levels of enzyme forms. The reversible connection results in a slope effect. The combination of an intercept effect and a slope effect means that P is a noncompetitive inhibitor of A when [B] is below saturation. When the fixed substrate, B, is saturating, the steady-state level of EA is reduced to zero, and so the reversible connection between E and EQ is broken. An intercept effect remains, but the slope effect is eliminated. The result is uncompetitive inhibition. When Q is used as the product inhibitor of A, the result is competitive inhibition in both cases. Q and A both bind to E, and rule 2 applies. When B is the variable substrate, P is a noncompetitive inhibitor, whether or not the fixed substrate is saturating. P binds to EQ and B binds to EA. These forms are different and reversibly connected. Saturation with A eliminates E, which is outside the connection. Q binds to E and B binds to EA. Again, the enzyme forms are different and reversibly connected. However, saturation with A eliminates the inhibition by Q since Q binds to E, whose steady-state level is reduced to zero.

All rules and notes except note 3 are illustrated in Table 2.1. Let us now consider note 3 and the elimination of intercept effects. The Theorell-Chance mechanism is a special version of the ordered mechanism:

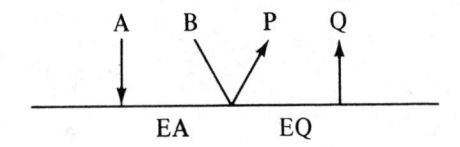

$$\begin{array}{cccc} A & B & P & Q \\ \downarrow & \searrow & \swarrow & \uparrow \\ \hline & EA & & EQ \end{array}$$

The product inhibition patterns differ from those of the ordered mechanism because of the middle step,

$$EA + B \rightleftharpoons EQ + P \qquad (101)$$

Referring to Table 2.2, consider the case of inhibition by P, variable substrate A, and saturating B. In the ordered case, these conditions give rise to an intercept effect and uncompetitive inhibition. P inhibits by converting EQ back to EA. This conversion is, of course, impossible when B is present in saturating amounts. In Eq. 101 infinite [B] overcomes finite [P] by pushing the reaction to the right. In the case of the random-rapid equilibrium pathway, all intercept effects are eliminated, in accordance with note 3. Competitive inhibition is the only type that remains. It is also important that saturation with either substrate pushes the equilibrium to the EAB complex and relieves the inhibition. In the case of a ran-

TABLE 2.2. Summary of Product Inhibition Patterns for $A + B \rightleftharpoons P + Q$

			Variable Substrate			
			A		B	
Pathway	Inhibitor		Unsaturated with B	Saturated with B	Unsaturated with A	Saturated with A

Ordered

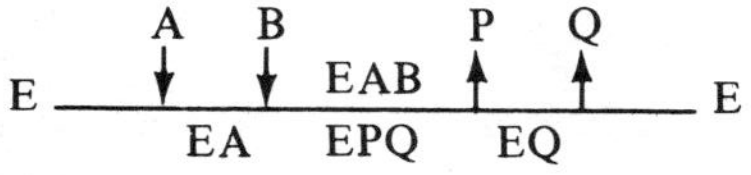	P	Noncompetitive	Uncompetitive	Noncompetitive	Noncompetitive
	Q	Competitive	Competitive	Noncompetitive	—

Theorell-Chance

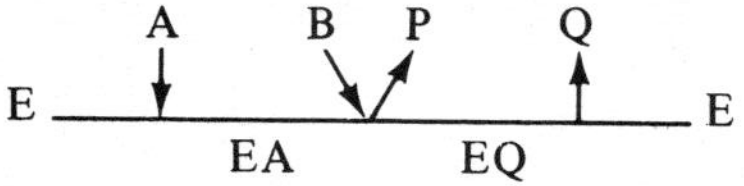	P	Noncompetitive	—	Competitive	Competitive
	Q	Competitive	Competitive	Noncompetitive	—

Random-Rapid Equilibrium

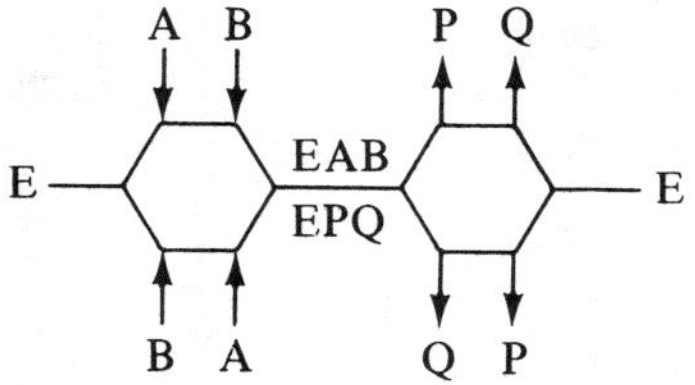	P	Competitive	—	Competitive	—
	Q	Competitive	—	Competitive	—

Ping-Pong

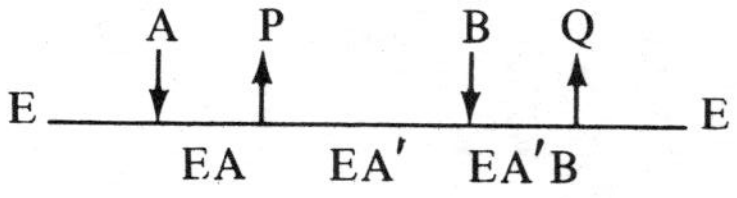	P	Noncompetitive	—	Competitive	Competitive
	Q	Competitive	Competitive	Noncompetitive	—

dom mechanism without the rapid equilibrium condition, inhibition would be noncompetitive in all cases.

The stepwise application of Cleland's rules is appropriate for termolecular reactions involving three substrates and three products and for other systems involving unequal numbers of substrates and products. In some instances, product can bind to an enzyme form to yield a nonproductive complex. In this case, the product inhibitor can actually serve simultaneously as a product inhibitor and a "dead-end" inhibitor. The important point here is that stepwise application of the rules and summation of the effects produced will handle these more complex situations.

The method of isotope exchange can provide valuable insights into the identification of the major pathway and alternative pathways, as well as kinetic constants. The derivation and application of rate equations for isotope exchange are treated by Plowman (16). We shall illustrate the utility of isotope exchange studies in verification of the ping-pong mechanism. Consider the reaction

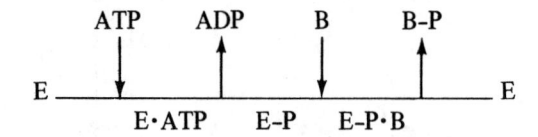

$$\begin{array}{ccccc} & ATP & ADP & B & B\text{-}P \\ & \downarrow & \uparrow & \downarrow & \uparrow \\ E & \underline{\qquad\qquad\qquad\qquad\qquad\qquad} & E \\ & E\cdot ATP & E\text{-}P & E\text{-}P\cdot B & \end{array}$$

If ATP were labelled with ^{14}C, the exchange of label between ATP and ADP should be catalyzed by the enzyme in the absence of the second substrate. The rate equation for the isotope exchange between ATP and ADP is

$$\text{rate} = \frac{C[ATP][ADP]}{C'[ATP] + C''[ADP] + [ATP][ADP]}$$

Variation of [ATP] at different fixed levels of ADP will yield double reciprocal plots with the parallel pattern typical of initial velocity results of the chemical reaction. Another application of isotope exchange studies is in the distinction between ordered and random pathways, which has been discussed in detail by Cleland (14).

G. COLLECTION AND TREATMENT OF ENZYME KINETICS DATA

The first actual step in a study of the kinetics of an enzyme-catalyzed reaction is the selection of an assay (17). A continuous assay is preferable to a "one-point" assay, in which [P] is determined at one time interval. In the continuous assay initial velocities can be determined with a recording spectrophotometer or pH-stat by measuring the slope of the initial linear part of the [P] versus time recording. In lieu of a convenient continuous assay, it is often wise to take the time to procure a new substrate or develop a coupled assay. Consider the enzyme asparaginase, which catalyzes the hydrolysis of the β-amide of asparagine:

$$\begin{array}{c} \text{COOH} \\ | \\ \text{CHNH}_2 \\ | \\ \text{CH}_2 \\ | \\ \text{CONH}_2 \end{array} \quad \xrightarrow{\text{asparaginase}} \quad \begin{array}{c} \text{COOH} \\ | \\ \text{CHNH}_2 \\ | \\ \text{CH}_2 \\ | \\ \text{COOH} \end{array} + \text{NH}_3$$

Asparagine   Aspartic acid

The rate may be established by sampling the reaction mixture at timed intervals and using a colorimetric technique for the determination of ammonia concentration in each sample. O'Leary and Mattes have described a continuous coupled assay that uses glutamate dehydrogenase (GDH) (18). The reaction is

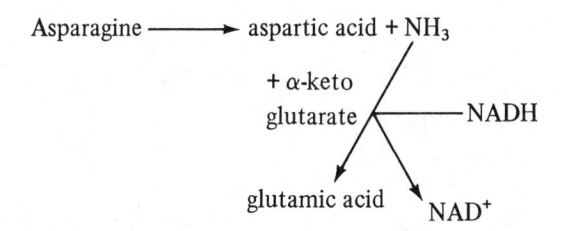

Asparagine ⟶ aspartic acid + NH$_3$
+ α-ketoglutarate ⟵ NADH
glutamic acid ⟵ NAD$^+$

The reaction can be followed conveniently by observation of the decrease in absorbance at 340 nm that accompanies the oxidation of NADH. The authors demonstrated that the observed rate showed a linear dependence on asparaginase concentration, but was independent of GDH concentration. Fluorescence methods based on the NADH/NAD$^+$ reaction are also valuable (17).

Along with photometric methods, the pH-stat finds wide use in enzymology. It is applicable to any enzyme reaction in which a proton is produced or consumed. The rate of automatic base or acid addition to maintain the desired pH is equal to the rate of reaction. The pH-stat can also be used in coupled assays if the first reaction can be linked to an enzyme-catalyzed reaction in which a proton is produced or consumed.

As one might expect, the quality of enzyme kinetic data varies with the assay technique and the enzyme system, so it is important to establish the magnitude of errors associated with kinetic constants. The Lineweaver-Burk plot is not so good as the Eadie plot for graphical determination of constants (17a). However, the double reciprocal plot is used because it predates the Eadie plot, inhibition patterns are more easily rationalized, and because kinetic constants are machine-calculated in most cases anyway. The quality of data at low substrate concentrations is usually poor. Considerable scatter is seen at high values of $1/S_0$ in Lineweaver-Burk plots. The procedure of Wilkinson weighs these points appropriately (19). Cleland has discussed the statistical treatment of reciprocal plots and has published computer programs that are used widely (20).

The selection of substrate range is important. An example is shown in Figure 2.19. A linear reciprocal plot is produced from data in the range where v_0 is nearly independent of S_0. Obviously, rates should be measured at S_0 values on both sides of K_m. Graphs of v_0 versus S_0 and $1/v_0$ versus $1/S_0$ should always be made before going to the computer terminal. Impure substrates, unstable enzymes, enzyme contaminants that catalyze transformation of the reactant or

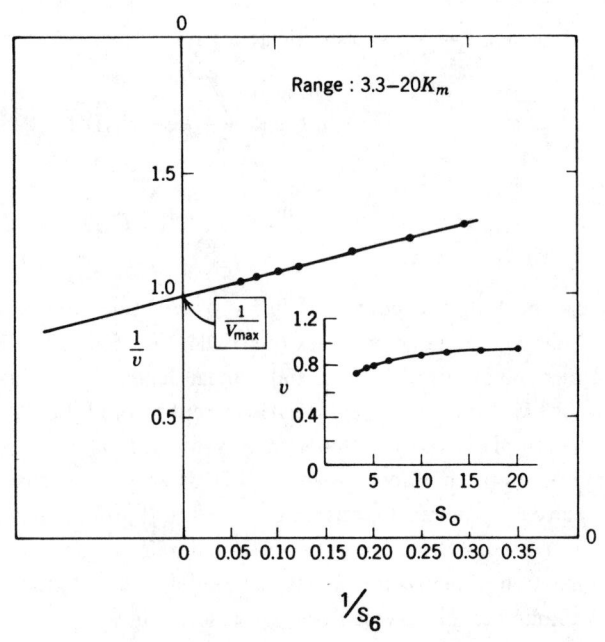

Figure 2.19. Linear double reciprocal. Notice that V_0 is nearly independent of S_0 (inset). From Segel, *Enzyme Kinetics*, Wiley, New York (1975), p. 47. Reproduced by permission.

product of interest, pH, ionic strength, and temperature are possible sources of error. These complications are more frequent and serious problems than the choice—or complete lack—of statistical massage.

REFERENCES

Specific

1. K. J. Laider, *Theories of Chemical Reaction Rates,* McGraw-Hill, New York (1969), p. 41.
2. V. Henri, *Acad. Sci., Paris* **135**, 916 (1902).
3. A. J. Brown, *Trans. Chem. Soc.* (Lond) **81**, 373 (1902).
4. L. Michaelis and M. L. Menten, *Biochem. Z.* **49**, 333 (1913).
5. H. J. Segel, in *The Enzymes,* Vol. 1, 2nd ed. (P. D. Boyer, H. Lardy, and K. Myrback, Eds.), Academic Press, New York (1959).
6. E. L. King and C. Altman, *J. Phys. Chem.* **60**, 1375 (1956). See also H. J. Fromm, *Biochem. Biophys. Res. Commun.* **40**, 692 (1970).
7. H. Lineweaver and D. Burk, *J. Am. Chem. Soc.* **56**, 658 (1934).
8. G. S. Eadie, *J. Biol. Chem.* **146**, 85 (1942).
9. M. F. Perutz, *Science* **201**, 1187 (1978).
10. J. I. Harris and J. E. Walker, in *Pyridine Nucleotide Dependent Dehydrogenase* (H. Sund, Ed.), de Gruyter, Berlin (1977).
11. G. Biesecker, J. I. Harris, J. C. Thierry, J. E. Walker, and A. J. Wonacott, *Nature* (London) **266**, 328 (1977).
12. D. Moras, K. W. Olsen, K. N. Sabesan, M. Buehner, G. C. Ford, and M. G. Rossman, *J. Biol. Chem.* **250**, 9137 (1975).

General References on Enzyme Kinetics

13. I. H. Segel, *Enzyme Kinetics,* John Wiley, New York (1975).
14. W. W. Cleland, in *The Enzymes,* Vol. 2, 3rd ed. (P. D. Boyer, Ed.), Academic Press, New York (1970), p. 1.
15. K. J. Laidler and P. S. Bunting, *The Chemical Kinetics of Enzyme Action,* Oxford, London (1973).
16. K. M. Plowman, *Enzyme Kinetics,* McGraw-Hill, New York (1976).

Collection and Treatment of Enzyme Kinetics Data

17. *Principles of Enzymatic Analysis* (H. U. Bergmeyer, Ed.), Verlag Chemie, New York (1978).
17a. J. E. Dowd and D. E. Riggs, *J. Biol. Chem.* **240**, 863 (1969).
18. M. H. O'Leary and S. L. Mattes, *Biochem. Biophys. Acta* **522**, 238 (1978).
19. G. N. Wilkinson, *Biochem J.* **80**, 325 (1961).
20. W. W. Cleland, *Adv. Enzymol.* **29**, 1 (1967).

H. EXERCISES

1. (a) The Henri-Michaelis-Menten equation may be written as

$$\frac{d[P]}{dt} = \frac{V_m(S_0 - [P])}{K_m + (S_0 - [P])}$$

Show that the integrated form is

$$\frac{[P]}{t} = \frac{-K_m}{t} \ln\left(\frac{S_0}{S_0 - [P]}\right) + V_m$$

(b) Suggest a plot for the determination of K_m and V_m.

2. Derive a rate equation for the following schemes.

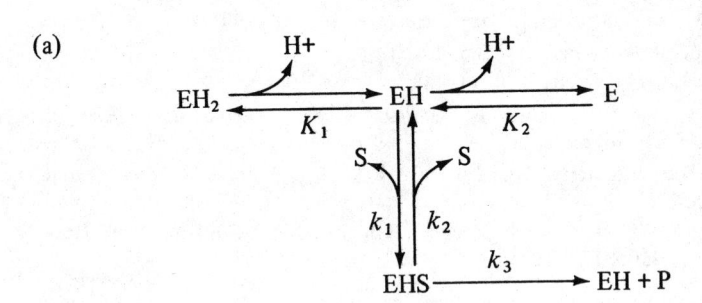

(a)

$$EH_2 \underset{K_1}{\overset{H^+}{\rightleftharpoons}} EH \underset{K_2}{\overset{H^+}{\rightleftharpoons}} E$$

$$EH + S \underset{k_2}{\overset{k_1}{\rightleftharpoons}} EHS \xrightarrow{k_3} EH + P$$

(b)

$$E + S \underset{K_m}{\rightleftharpoons} ES \xrightarrow{k_{cat}} E + P$$

$$E + I \underset{K_I}{\rightleftharpoons} EI$$

$$ES + I \underset{K_I'}{\rightleftharpoons} ESI$$

(c)

$$E + S \underset{K_m}{\rightleftharpoons} ES \xrightarrow{k_{cat}} E + P$$

$$E \longrightarrow E^*(\text{inactive})$$ (assume inactivation of free E is first order)

3. Prove that the coordinates of the point of intersection of the plot in Figure 2.18a are equal to $-K_{i,a}$ and $1/V_m(1-K_{m,a}/K_{i,m})$
4. Using Cleland's rules, construct a table similar to Table 2.1 for the ping-pong pathway (A + B ⇌ P + Q).

CHAPTER THREE

STRUCTURE OF THE ACTIVE CENTER: AMINO ACID SIDE CHAINS, COENZYMES, AND METAL IONS

Kinetics studies dominated enzyme research until effective methods were developed for protein structure determination. It was Sanger's work on sequencing and the development of the amino acid analyzer in the laboratory of Moore and Stein that led to the knowledge of covalent structures of several proteins in the early 1960s. Chemical modification studies and X-ray crystallography soon revealed important residues at the catalytic and binding sites of enzymes. The realization that enzymes are homogeneous substances with definable structure led many workers to investigate the reactions of these proteins with other chemicals in the hope of correlating discrete structural elements with function. Approaches taken in such studies include general chemical modification, use of pseudosubstrates, trapping of covalent intermediates with an exogenous reagent, and affinity labeling. X-ray crystallography has provided valuable confirmation of results from chemical experiments in solution and evidence for additional active site residues not found by chemical studies.

Table 3.1 shows the constituents of enzyme active sites that are important in the catalytic process.

TABLE 3.1. Catalytic Constituents of Enzymatic Active Sites[a]

Coenzymic Catalysts or Reactants

Oxidation-Reduction Systems
  Nicotinamide-adenine dinucleotide (NAD)
  Nicotinamide adenine-dinucleotide phosphate (NADP)
  Flavin nucleotides such as flavin mononucleotide (FMN) and flavin adenine dinucleotide (FAD)
  Metal porphyrin complexes such as those found in the cytochromes, cobamide, peroxidase, and catalase
  Ascorbic acid
  Lipoic acid (thioctic acid)
  Coenzyme Q (ubiquinone)
  Metal ions such as Cu^{2+}, Fe^{2+}
Nonoxidation-Reduction Systems
  Thiamine pyrophosphate
  Pyridoxal phosphate
  Folic acid (pteroyl-L-glutamic acid)
  Biotin
  Glutathione
  S-adenosylmethionine
  Coenzyme A
  Adenosine monophosphate, diphosphate, and triphosphate (AMP, ADP, ATP)
  Uridine phosphate
  Various metal ions, mainly Zn^{2+}, Mn^{2+}, Mg^{2+}, Cu^{2+}, Co^{2+}
  $4'$-phosphopantetheine
  Pyruvate

Constituents of the Protein

Carboxylate ion
Alcoholic hydroxyl group
Phenolic hydroxyl group
Ammonium ion (amine)
Imidazolium ion (imidazole)
Guanidinium ion
Indole ring
SH group
SCH_3 group
Peptide or amide group

[a]From M. L. Bender, in *Encyclopedia of Polymer Science and Technology*, Vol. 6, John Wiley and Sons, New York (1967), p. 1.

A. AMINO ACID SIDE CHAINS

Chemical Modification of Enzymes with Nonspecific Reagents

Means and Feeney have compiled a formidable list of general reagents for proteins (1). The use of nonspecific reagents can be valuable if selection of the reaction conditions results in restriction of the reactivity. Unusual reactivity of groups at the catalytic site may be important in this regard. Consider the electrophilic reagent iodoacetic acid, ICH_2COOH, an activated alkyl halide capable of reacting with many groups on proteins, especially thiol groups, primary amines, phenolate groups, and imidazole groups. In the case of the enzyme ribonuclease, Crestfield, Stein, and Moore were able to limit the reactivity of iodoacetate to histidine by buffering the reaction medium at pH 5.5 (2, 3), They found two alkylated products, each with one modified histidine. One derivative was found to be totally inactive; the other retained only seven percent of the initial activity. With the aid of the primary sequence, Crestfield, Stein, and Moore were able to conclude that two histidine residues, at positions 12 and 119, were essential for full activity and were sufficiently close in the three-dimensional structure that reaction of one prevented reaction of the other.

A useful approach with general reagents is differential labeling in the presence and absence of a competitive inhibitor (Fig. 3.1). For example, Eyl and Inagami established that the β-carboxyl group of Asp-177 was the anionic binding group at the active center of trypsin (4). These workers derivitized the carboxyl groups of trypsin with water-soluble carbodiimide and glycinamide in the presence of benzamidine, a potent competitive inhibitor of the enzyme. Water-soluble carbodiimide and glycinamide react as follows:

$$\text{Protein-COOH} + \text{R-N=C=N-R}' \longrightarrow \text{Protein-}\underset{\underset{\displaystyle O}{\|}}{C}\text{-O-}\underset{\underset{\displaystyle +NH-R'}{|}}{\overset{\overset{\displaystyle R}{|}}{\underset{\|}{C}}}\text{NH}$$

$$\overset{NH_2CH_2CONH_2}{\swarrow}$$

$$\text{Protein-}\underset{\underset{\displaystyle}{}}{\overset{\overset{\displaystyle O}{\|}}{C}}\text{-NHCH}_2\text{COHN}_2$$

After the initial reaction step, benzamidine was removed by dialysis and the reaction was repeated, using ^{14}C-labeled glycinamide. The labeled enzyme was sub-

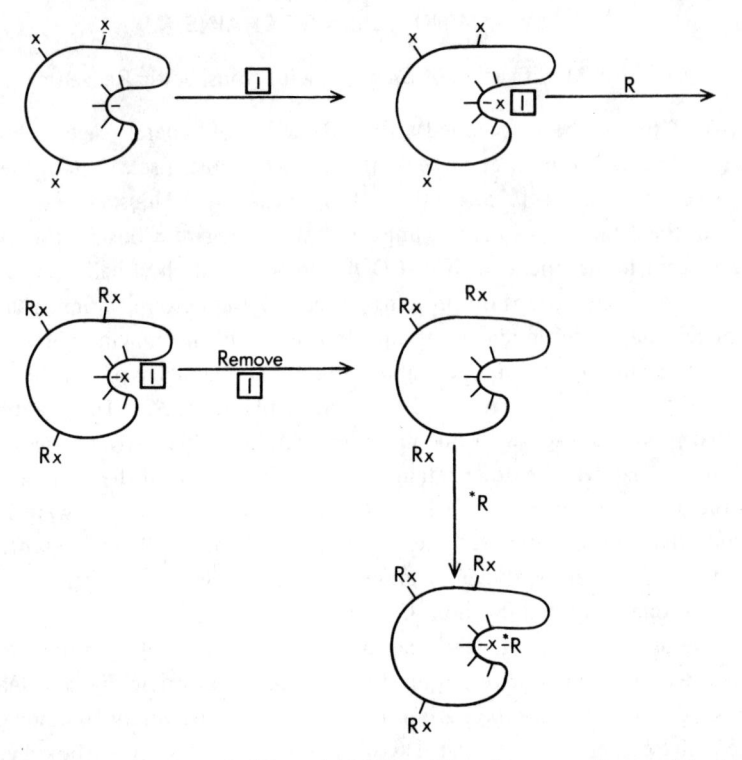

Figure 3.1. Schematic view of differential labeling. First the enzyme is reacted in the presence of an inhibitor that protects the active site. After removal of the inhibitor, the general reagent (now radioactive) is used a second time to label the active site residue.

jected to structural analysis, and the result was that Asp-177 was found to be amidated with glycinamide. This result was confirmed by crystallographic studies of Stroud and coworkers (5).

Pseudosubstrates

Acetylcholinesterase and serine proteases, including chymotrypsin and trypsin, are inactivated in a stoichiometric reaction with diisopropylphosphofluoridate (DFP). It is now known that these hydrolases have serine at the active center. This serine residue possesses unusual reactivity toward DFP because of the enhanced nucleophilicity of the hydroxyl group and the attraction of the apolar isopropyl groups to the binding site. For α-chymotrypsin the reaction may be illustrated as follows:

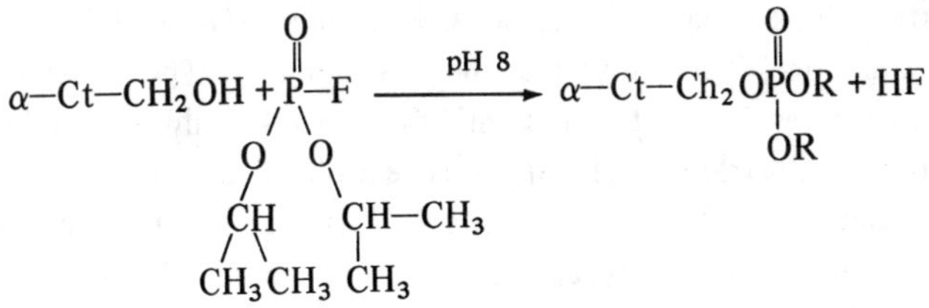

A. AMINO ACID SIDE CHAINS

$$\alpha\text{--Ct--CH}_2\text{OH} + \underset{\underset{\underset{\text{CH}_3\text{CH}_3}{|}}{\overset{\overset{\text{CH}}{/\backslash}}{O}}}{\overset{\overset{O}{\|}}{\underset{}{P}}-F}\xrightarrow{\text{pH 8}} \alpha\text{--Ct--Ch}_2\underset{\underset{OR}{|}}{\overset{\overset{O}{\|}}{O-P-OR}} + HF$$

DFP   DIP–Ct

DFP has been termed a pseudosubstrate because a covalent ester intermediate analogous to acyl intermediates that result with specific substrates is formed. The important difference is that the phosphoryl serine derivative turns over very slowly.

Hartly and Kilby found that α-chymotrypsin catalyzes the hydrolysis of p-nitrophenyl acetate by way of a double displacement pathway (6):

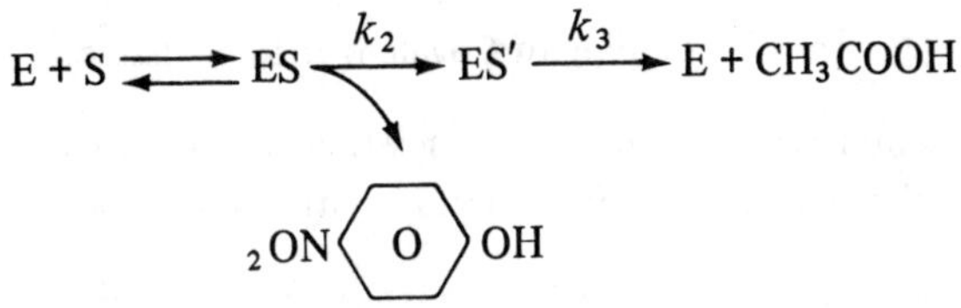

$$E + S \rightleftharpoons ES \xrightarrow{k_2} ES' \xrightarrow{k_3} E + CH_3COOH$$

$$\downarrow$$

$$_2ON\langle O\rangle OH$$

When the reaction is carried out at pH 5, ES' (the acylenzyme) turns over very slowly. Oosterbaan and coworkers reacted chmyotrypsin with (^{14}C)p-nitrophenyl acetate and isolated a ^{14}C-acetyl peptide. The acetyl group was found to be esterified to the same serine residue that reacts with DFP (7).

Glyceraldehyde 3-phosphate dehydrogenase catalyzes the formation of 1,3-phosphoglycerate by means of oxidative phosphorylation of glyceraldehyde 3-phosphate:

$$\begin{array}{l}\text{CHO}\\|\\\text{CHOH} + NAD^+ + HPO_3^{2-}\\|\\\text{CH}_2\text{O}\\\quad|\\\quad PO_3^{2-}\end{array} \longrightarrow \begin{array}{l}\overset{\overset{O}{\|}}{C}-O-PO_3^{2-}\\|\\\text{CHOH} + NADH + H^+\\|\\\text{CH}_2-O\\\qquad|\\\qquad PO_3^{2-}\end{array}$$

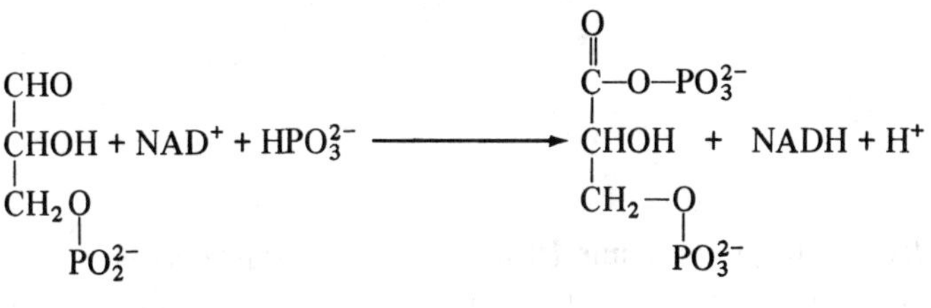

This enzyme also reacts with p-nitrophenyl acetate to form an acyl enzyme that is not active with the natural substrates. When labeled *p*-nitrophenyl acetate was used, evidence for ^{14}C-*S*-acetyl cysteine was found (8). The cysteine residue was identical to the residue that reacts with thiol-specific inhibitors of the enzyme. Another pseudosubstrate for glyceraldehyde 3-phosphate dehydrogenase was discovered in Bernhard's laboratory (9). β-2 (furyl) acryloyl phosphate, **1**, reacts with the active-site thiol of glyceraldehyde-3-phosphate dehydrogenase to form a stable furyl-acryloyl derivative with an absorption maximum at 344 nm. The reagent is superior to *p*-nitrophenyl acetate because the intermediate can be detected spectrophotometrically.

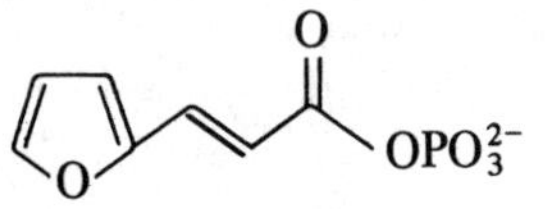

$$\text{furyl-CH=CH-C(=O)-OPO}_3^{2-}$$

1

Trapping of Covalent Intermediates

Many transaminases and other enzymes that use pyridoxal-5-phosphate can be labeled at the active site by reduction with sodium borohydride under mild conditions:

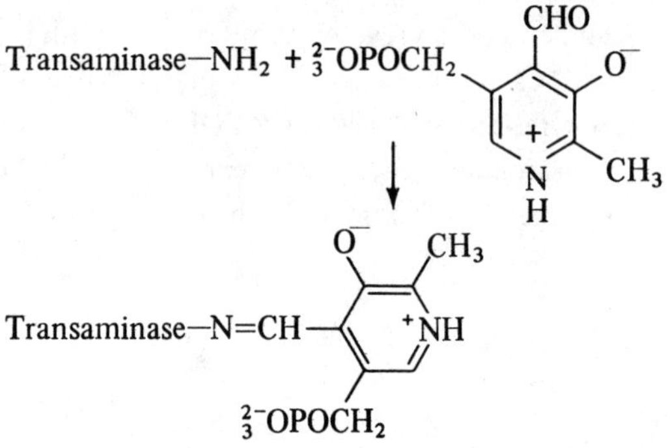

$$\text{Transaminase-NH}_2 + {}_3^2{}^-\text{OPOCH}_2\text{-[pyridoxal]}$$

$$\downarrow$$

$$\text{Transaminase-N=CH-[pyridoxamine-P]}$$

Borohydride reduces the imine (Schiff base) to a stable alkylamine. Other enzymes having imine intermediates along the reaction pathway can be similarly labeled. Aldolase is a good example:

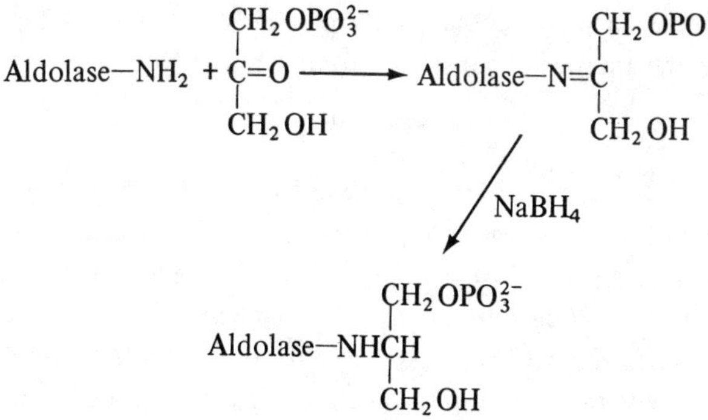

A. AMINO ACID SIDE CHAINS

$$\text{Aldolase-NH}_2 + \underset{\underset{\text{CH}_2\text{OH}}{|}}{\overset{\overset{\text{CH}_2\text{OPO}_3^{2-}}{|}}{\text{C}=\text{O}}} \longrightarrow \text{Aldolase-N}=\underset{\underset{\text{CH}_2\text{OH}}{|}}{\overset{\overset{\text{CH}_2\text{OPO}_3^{2-}}{|}}{\text{C}}}$$

$$\downarrow \text{NaBH}_4$$

$$\text{Aldolase-NH}\underset{\underset{\text{CH}_2\text{OH}}{|}}{\overset{\overset{\text{CH}_2\text{OPO}_3^{2-}}{|}}{\text{CH}}}$$

Dihydroxyacetone phosphate forms a Schiff base intermediate with a lysine at the active center. Reduction with borohydride results in the specific labeling of the active sites of a number of aldolase enzymes (8). Shemin and coworkers have used this trapping method in studies of δ-aminolevulinic acid dehydratase, which catalyzes the formation of porphobilinogen from two molecules of δ-aminolevulinic acid:

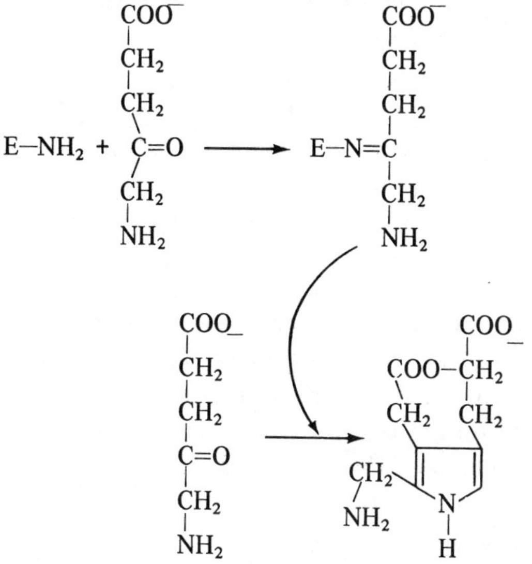

When the enzyme was treated with NaBH₄ in the presence of δ-aminolevulinic acid $-4^{14}C$, simultaneous inactivation and labeling occurred. Treatment with borohydride in the absence of substrate produced no effect.

Affinity Labeling

The selectivity and strength of protein-small molecule interactions are widely accepted as important and distinctive phenomena of biological systems. As described earlier, the remarkable specificity of enzymes, antibodies, and permeases is a fascinating and fundamental part of biochemical science. Subdisciplines stemming from the biospecific reactions of enzymes and other biopolymers have evolved. These fields include affinity labeling, affinity chromatography, affinity cytochemistry, affinity therapy, immunological analysis, and targeted drug delivery. In a limited definition, affinity labeling is the design, synthesis, and reaction of a substrate analog. The product must be a stable covalent complex of enzyme and inhibitor. Structural analysis of the labeled enzyme yields a contituent amino acid side chain at the active center. The earliest mention of affinity labeling that I have found is by Baker and coworkers in 1959 (10), and its first demonstration, as we now view it, was the reaction of 4-(iodoacetamido)-salicylic acid (2) with glutamate dehydrogenase (11). The design of 2 was based on the finding that salicylic acid (3) reversibly inhibits glutamate dehydrogenase.

Reaction of 2 with the enzyme was much faster than the reaction of enzyme with iodocetamide and was retarded by known reversible inhibitors; also, the rate of reaction was dependent on the K_I's of various compounds related to 2. These findings suggest the formation of a noncovalent enzyme-inhibitor complex prior to the alkylation of a nearby amino acid side chain:

$$E\text{-}H + R\text{-}X \underset{k_{-1}}{\overset{k_1}{\rightleftharpoons}} E\text{-}H \cdot RX \overset{k_2}{\longrightarrow} E\text{-}R + HX$$
(affinity label)

Reversible complexation of the affinity label by the enzyme would mean that saturation kinetics would be observed for the inactivation reaction—that is, the

A. AMINO ACID SIDE CHAINS

observed rate of reaction should show a hyperbolic dependence on the concentration of label. The rate equation for the inactivation of E by RX is:

$$v_{inact} = \frac{V_{inact}[RX]}{(k_{-1} + k_2)/k_1 + [RX]} \tag{1}$$

V_{inact} is the rate of inactivation at infinite [RX], or the rate when E = 0 (11a).

Affinity labeling of enzyme active centers was stimulated by work on halomethylketones of N-blocked amino acids in Shaw's laboratory. TPCK (L-1-chloro-3-tosylamido-4-phenyl-2-butanone), 4, reacts stoichiometrically with α-chymotrypsin (12). The inhibitor closely resembles specific substrates, such as N-benzoyl-L-tyrosine ethyl ester (5):

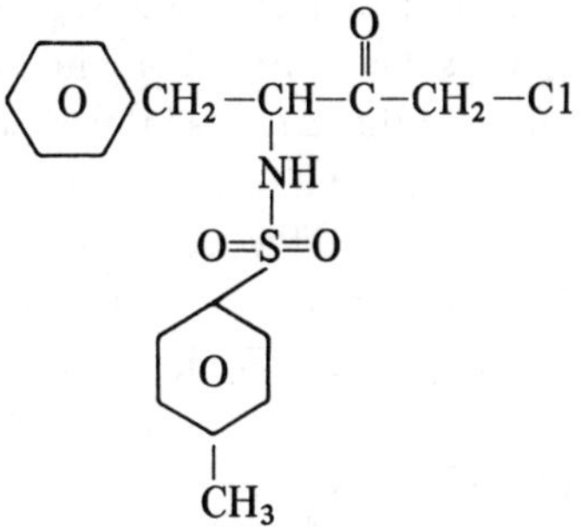

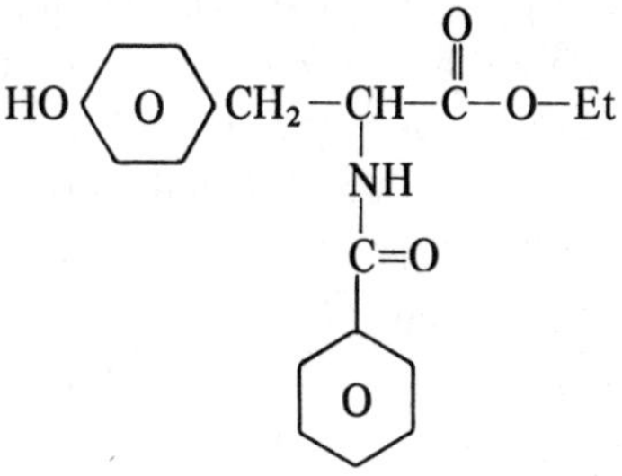

Reaction of TPCK with chymotrypsin shows a pH optimum of 6.8 (14). Loss of one histidine was demonstrated by amino acid analysis. Pepsin treatment followed by chromatography yielded a peptide containing alkylated histidine-57 (15).

Similar studies were done with trypsin, using a positively charged inhibitor, TLCK (6)

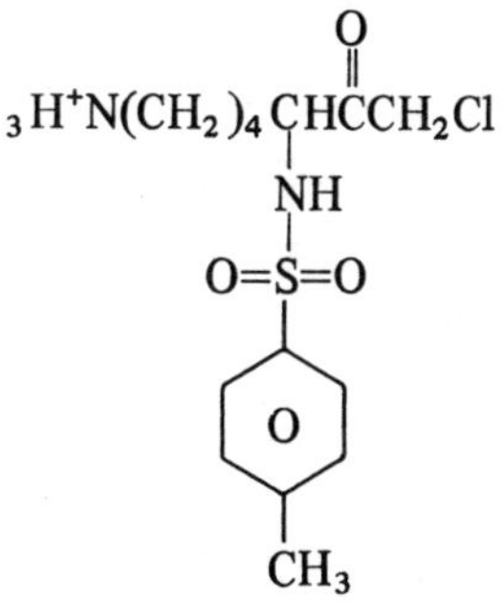

Compound **6** was found to be a true affinity label in that chymotrypsin was unaffected by its presence. The reaction with trypsin was prevented by a competitive inhibitor, benzamidine, and also by 8 M urea. Halomethylketones have been used to label a wide variety of enzymes other than serine proteases. Valyl-*t*RNA synthetase is inhibited by L-3-amino-1-chloro-4-methylpentane-3-1, **7**, which is a chloromethylketone analog of L-valine, **8**,

$$\underset{7}{\underset{NH_2}{\overset{CH_3}{CH_3CHCHCCH_2Cl}}} \qquad \underset{8}{\underset{NH_2}{\overset{CH_3}{CH_3CHCHCOH}}}$$

Valine protects the enzyme from inactivation by **7**. A plot of the reciprocal of the observed rate constant for inactivation versus the reciprocal of inhibitor concentration was found to be linear, which suggests that a reversible complex precedes the reaction of the inhibitor with a vital component of the active site.

In 1962 Singh, Thornton, and Westheimer introduced the technique of photoaffinity labeling (16). In this approach a label that contains a center capable of undergoing a photochemical reaction is made. The product of the photochemical activation can subsequently react with a group at the active center. Carbenes and nitrenes are frequently the photoactivated intermediates. The first photoaffinity label was *p*-nitrophenyl diazoacetate, **9**.

$$\underset{9}{_2ON{-}\langle O \rangle{-}O{-}\overset{O}{\overset{\|}{C}}{-}CH_2N_2} \xrightarrow{\alpha\text{-ct}} \alpha{-}Ct{-}CH_2O{-}\overset{O}{\overset{\|}{C}}{-}CH_2N_2$$

$$\downarrow h\nu$$

$$\alpha{-}CtCH_2O\overset{O}{\overset{\|}{C}}{-}CH{:} \quad \mathbf{10}$$

$$\overset{H_2O}{\swarrow} \qquad \downarrow$$

$$\alpha{-}Ct{-}CH_2O\overset{O}{\overset{\|}{C}}CH_2OH \qquad \text{labeling reaction}$$

$$\mathbf{11}$$

$$\alpha{-}Ct{-}\begin{bmatrix} {-}X{\diagdown} \\ \\ {-}CH_2O{\diagup} \end{bmatrix} \overset{CH_2}{\underset{\overset{\|}{O}}{C}}$$

12

A. AMINO ACID SIDE CHAINS

Chymotrypsin reacts with **9** to give a diazoacetyl intermediate. Photolysis yields the carbene, **10**, which reacts either with water, to give the glycolic acid derivative that turns over, or with a group at the active center. Acid hydrolysis and analysis showed labeling at serine (by rearrangement of **10**), histidine, and tyrosine near the active-site serine (17,18).

Bayley and Knowles pointed out some disadvantages of photogenerated carbene intermediates (19). Nitrenes are similar in reactivity to carbenes, but are somewhat more discriminate. For example, the photogenerated nitrene would have a greater tendency to react with an OH group than with a C–H bond. Aryl azides are the major source of nitrenes:

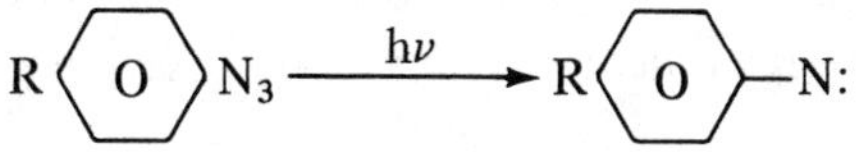

$$R\langle O \rangle N_3 \xrightarrow{h\nu} R\langle O \rangle -N:$$

For aromatic ligands, an azido group can be incorporated in place of an atom or group on the aromatic nucleus. Bridges and Knowles (20) probed the specificity pocket of α-chymotrypsin by photolysis of the 4-azidocinnamoyl derivative, **13**.

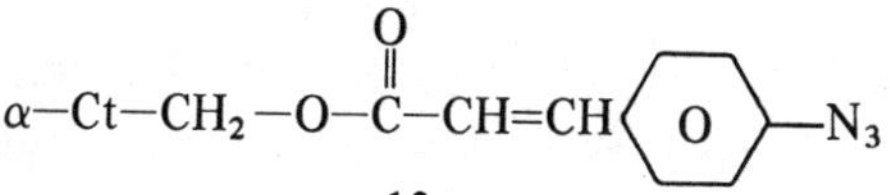

$$\alpha-Ct-CH_2-O-\overset{O}{\underset{\|}{C}}-CH=CH\langle O \rangle -N_3$$
$$\mathbf{13}$$

Sonnenberg and coworkers made photoreactable derivatives of Phe-*t*RNA to label the ribosomal peptidyltransferase center (21). The synthesis of *p*-azido (*N-t-Boc*)phenylalanyl-Phe-*t*RNA follows.

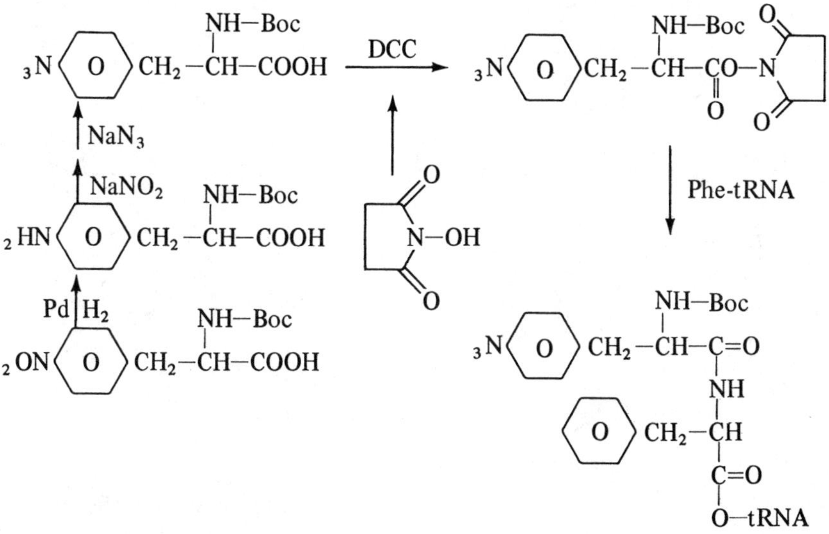

The azido derivative of *Phe-t*RNA binds reversibly to 70*S* ribosomes in the presence of poly(U). Irradiation brings about a reaction that results in the attachment of the label to the 23*S* RNA of the 50*S* subunit.

Another group of affinity labels includes the mechanism-based irreversible inhibitors (suicide substrates) (22). In this case the substrate is unreactive. A reactive product is formed as a result of an enzyme-catalyzed reaction:

$$E + S \rightleftharpoons ES \longrightarrow E \cdot I \longrightarrow E\text{-}I$$

The specificity of mechanism-based inhibitors is attractive. Very reactive labels are also generated *in situ*, which would not be useful in the normal labeling mode. Compound **14** is a potent inhibitor of aspartate aminotransferase.

$$CH_3O\diagdown CH\diagdown CO_2H$$
$$CH\ \ CH$$
$$|$$
$$NH_2$$

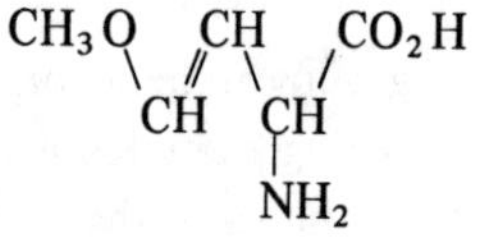

14

When the α-amino group reacts with enzyme-bound pyridoxal, a potent electrophilic center is generated; it reacts with a nucleophilic group (N) at the enzyme active center:

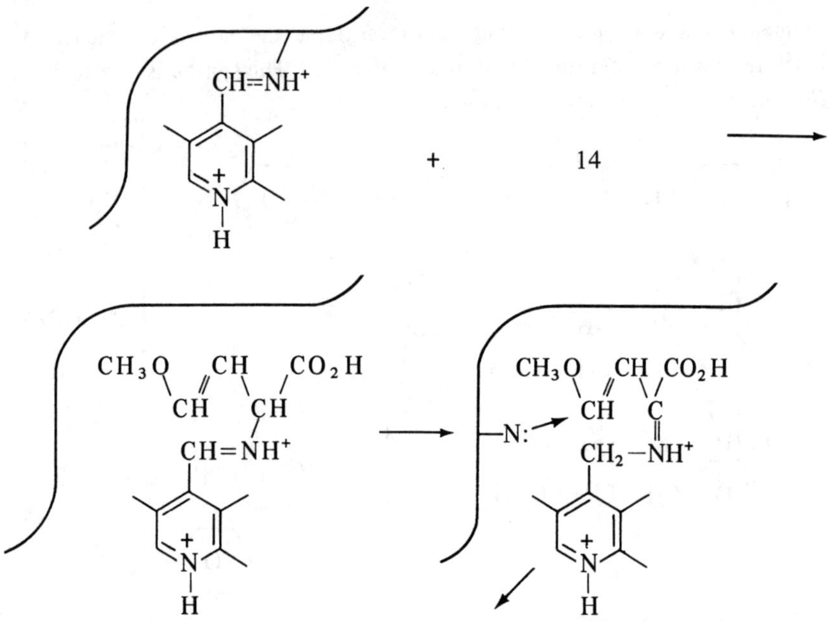

labeled aspartate aminotransferase

X-Ray Crystallography

Protein crystallography provides approximate pictures of empty enzymes, enzyme-coenzyme complexes, and nonproductive enzyme inhibitor complexes. The argument over whether or not these crystal structures resemble the solution structures will not be addressed here. Suffice it to say that the cooperation of crystallographers and "solution-phase" enzymologists has yielded much valuable information on the structure and function of protein active centers.

An important discovery of an unsuspected amino acid side chain at an enzyme active center was made by Blow and collaborators (23). The crystal structure of α-chymotrypsin revealed that the hydroxyl oxygen of Ser-195 was 3 Å removed from N^3 of His-57. This finding was precisely that predicted by chemical and kinetics studies. However, it was also found that the carboxylate group of Asp-102 was aligned nicely with the hydrogen of N^1 of His-57 (Fig. 3.2). The Asp-His-Ser triad was subsequently found in other serine proteases, including trypsin, elastase, *S. griseus* protease B, α-lytic protease, and subtilisn. The alignment of the Asp-His-Ser triad led to the proposal of the "charge-relay" mechanism, discussed in detail in Chapter 4.

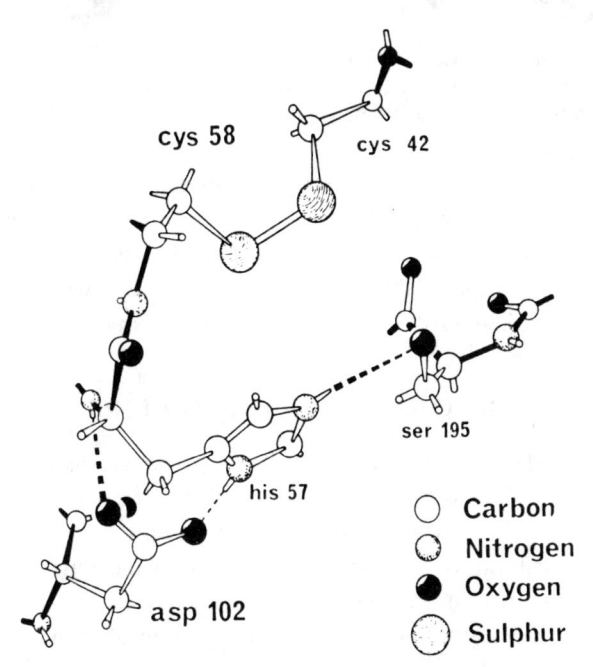

Figure 3.2. The catalytic triad of α-chymotrypsin. From Blow et al., *Nature* **221**, 337 (1969). Reproduced by permission.

104　　　　　　　STRUCTURE OF THE ACTIVE CENTER

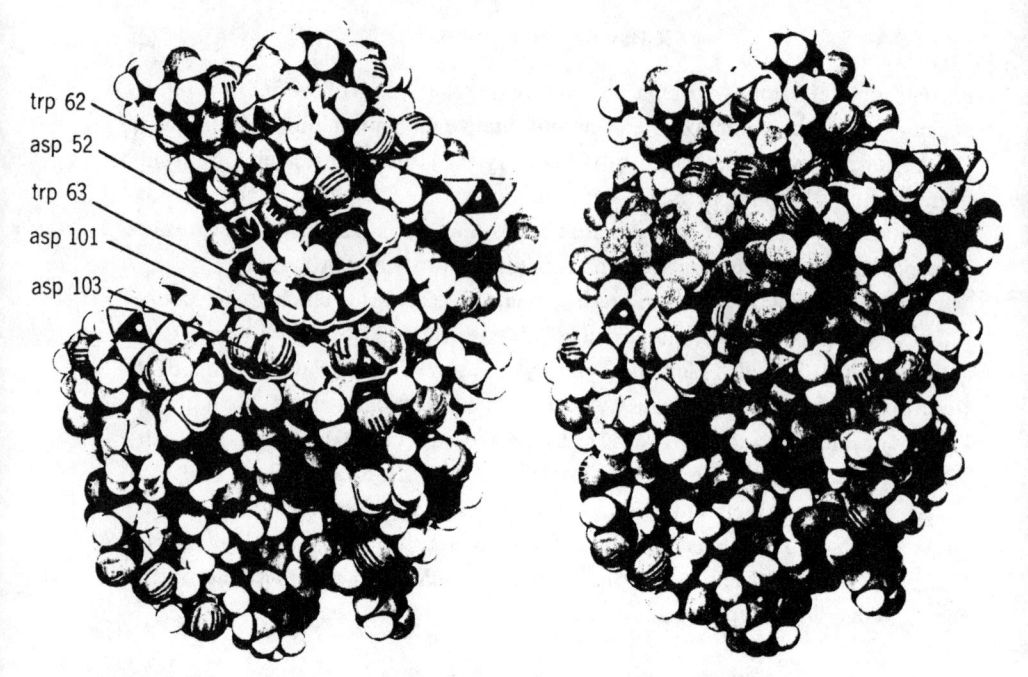

trp 62
asp 52
trp 63
asp 101
asp 103

Figure 3.3. The binding crevice of lysozyme with and without the hexasaccharide substrate. From R. E. Dickerson and I. Geiss, *The Structure and Action of Proteins,* Harper and Row, New York (1969), p. 75. Reproduced by permission.

There are numerous other examples confirming the presence of essential groups and discoveries of previously unsuspected active site components. Crystallographic pictures have also been extremely useful in the definition of binding sites. A good illustration of this can be seen in Figure 3.3, which depicts the enzyme lysozyme with and without a hexasaccharide substrate.

B. COENZYMES AND COFACTORS

The terms *coenzyme, prosthetic group,* and *cofactor* are used to describe the non-protein part of the enzyme active center. Rather than discuss vague distinctions among the definitions of these terms, it can simply be said that these molecules are not made up of the normal amino acid components of proteins. The coenzyme may or may not be regenerated in the reaction catalyzed by the enzyme with which it associates; in this case the additional term *cosubstrate* may be used.

B. COENZYMES AND COFACTORS

The presence of a coenzyme at the active center can often be demonstrated by separating the protein from the coenzyme by dialysis, ammonium sulfate treatment, extraction with organic solvent, and other such methods. The inactive "apoenzyme" can then be recombined with the pure coenzyme to form an active "holoenzyme." Association of coenzyme with protein can be followed by activity measurements, and in some cases by absorption changes in the UV or visible regions. The spectrophotometric "titration" of glyceraldehyde-3-phosphate dehydrogenase with NAD is illustrated in Figure 3.4. The binding of NAD by glyceraldehyde-3-phosphate-dehydrogenase is accompanied by a change in absorption at 360 nm. An end point is reached in this case because the binding of NAD by glyceraldehyde-3-phosphate-dehydrogenase is extremely strong ($K_d = 10^{-6}$ M). In cases where binding is not particularly strong ($K_d \geqslant 10^{-3}$ M), this titration approach does not work, but equilibrium dialysis can be used to obtain the number of binding sites per molecule of enzyme.

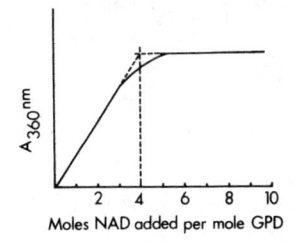

Figure 3.4. The spectrophotometric "titration" of glyceraldehyde-3-phosphate dehydrogenase. Addition of NAD results in a change in absorption to the point at which all sites are occupied.

An X-ray crystallography picture of a complex of NAD$^+$, glyceraldehyde-3-phosphate dehydrogenase appears in Figure 3.5 (25). The major driving force for binding is probably provided by the interaction of the adenine moiety with the aromatic rings of Phe-34 and Phe-99. A similar hydrophobic interaction is evident in the case of the *Bacillus stearothermophilus* enzyme, which involves the adenine ring of the NAD and side chains of Phe-99 and Leu-33 (26). A more recent structure from Rossman's laboratory suggests that the ϵ-ammonium ion of Lys-183 is not in a direct ionic linkage with phosphate of NAD$^+$, at least in the case of the lobster muscle enzyme.

The crystallographic studies of NAD-linked dehydrogenases have revealed some important structural relationships. The X-ray structures of lactate dehydrogenase (LDH), soluble malate dehydrogenase (*S*-MDH), liver alcohol dehydrogenase (LADH), and glyceraldehyde-3-phosphate dehydrogenase (GAPDH) are available (25). These enzymes all possess distinct catalytic domains and

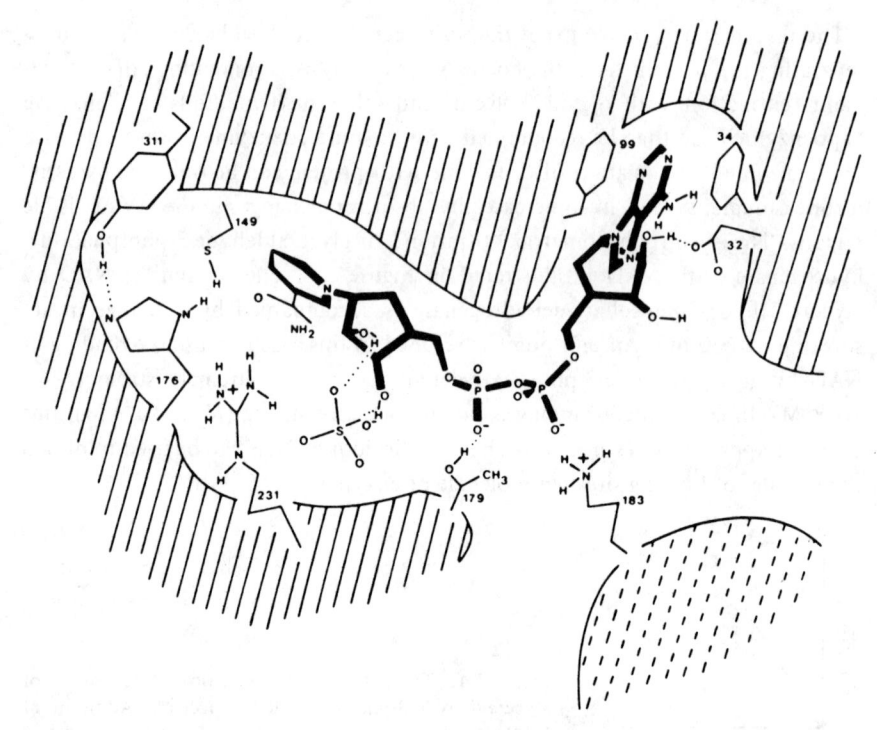

Figure 3.5. The binding of NAD⁺ to glyceraldehyde-3-phosphate dehydrogenase. From Rossman, *The Enzymes* **11**, 61 (1975). Reproduced by permission.

dinucleotide-binding domains. The catalytic domains of LDH and *S*-MDH are similar, but they do not resemble the catalytic domains of LADH and GAPDH. The dinucleotide-binding domains of all four enzymes are very similar. The NAD⁺-binding domain consists of a six-stranded parallel β-sheet flanked by helices. This arrangement of β-strands (arrows) and helices (cylinders) is shown in Figure 3.6 (27). The similarity of the dinucleotide-binding domains and the differences in the catalytic domains have formed the basis for the hypothesis that the dehydrogenases have evolved from the fusion of a gene coding for the dinucleotide domain with a series of genes, each coding for a catalytic domain.

Nicotinamide and flavin nucleotides are examples of coenzymes that are reversibly bound to enzyme active centers in a noncovalent manner. Other coenzymes are covalently attached. Pyridoxal phosphate is bound to the active centers of a number of enzymes through a Schiff base (aldimine) linkage. In reactions of amino acids, a "transaldimation" reaction, which breaks the covalent link be-

B. COENZYMES AND COFACTORS

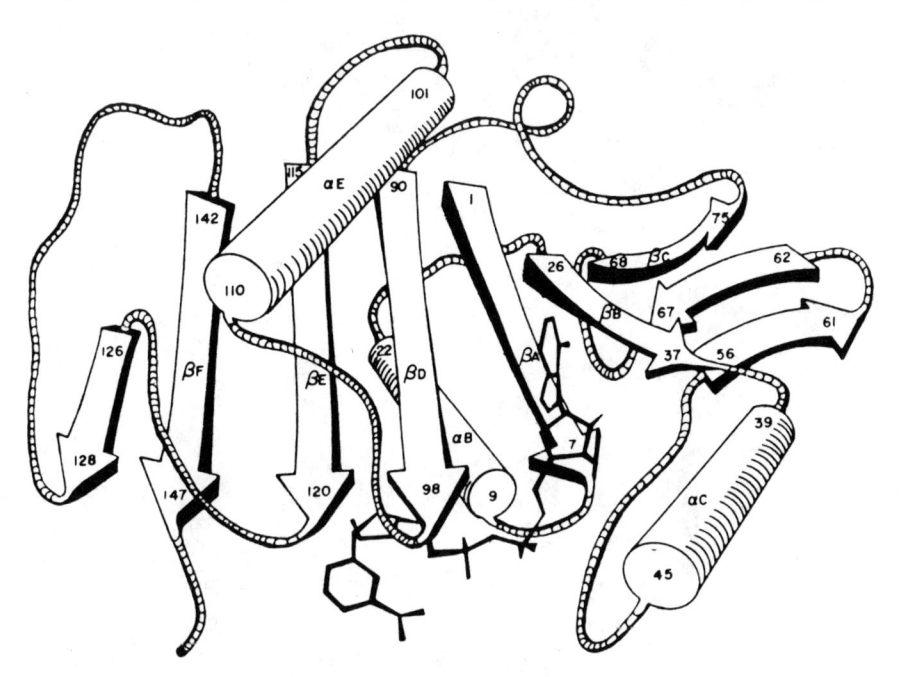

Figure 3.6. The NAD binding domain of dehydrogenases. From J. I. Harris and M. Waters, *Enzymes*, **11**, 12 (1975), reproduced by permission.

tween enzyme and coenzyme, occurs. However, the pyridoxal derivative remains bound by noncovalent attractions (Fig. 3.7). The attachment of pyridoxal phosphate can be established by reduction of the Schiff base with NaBH$_4$.

Stable amide linkages occur between lysine residues and the carboxyl groups of lipoic acid, **15**, and biotin, **16**.

$$\begin{array}{cc}
\underset{15}{\overset{\displaystyle CH_2\!\!-\!\!CH_2}{\underset{S-S}{|}}CH(CH_2)_4\overset{O}{\overset{\|}{C}}NH(CH_2)_4-ENZYME} & \underset{16}{\overset{\displaystyle HN\overset{\overset{O}{\|}}{\overset{C}{\diagdown}}NH}{\underset{S}{|}}HC\!-\!CH\;CH_2\;CH(CH_2)_4\overset{O}{\overset{\|}{C}}NH(CH_2)_4-ENZYME}
\end{array}$$

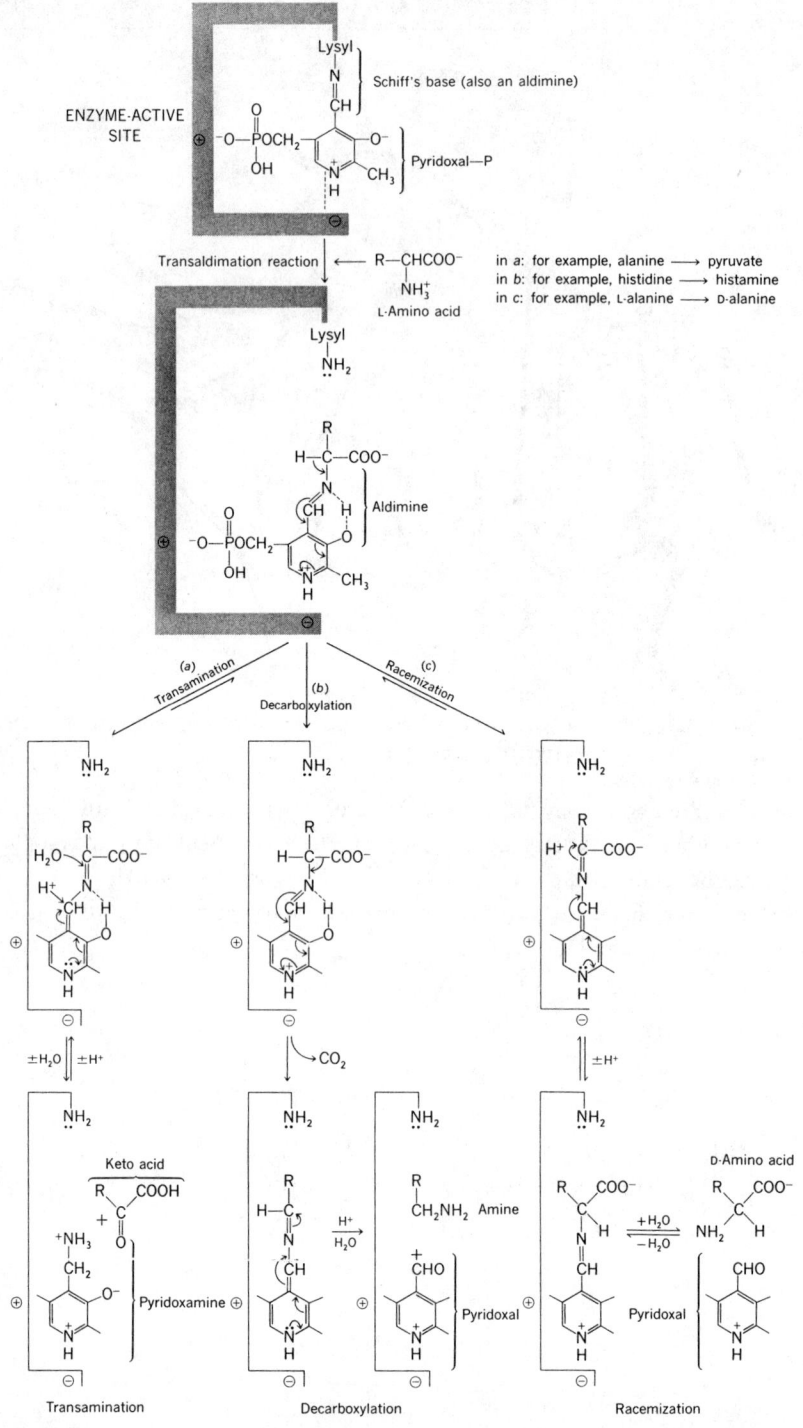

Figure 3.7. Enzyme-bound pyridoxal phosphate. From Conn and Stumpf, *Chemistry of Biological Compounds,* 3rd ed., Wiley, New York (1972), p. 216. Reproduced by permission.

Lipoyl enzymes are components of the pyruvate dehydrogenase complex and the α-ketoglutarate dehydrogenase complex. Lipoic acid is a cofactor in the generation and transfer of acyl groups. Biotin-containing enzymes are involved in carboxylation reactions. Isotopically labeled intermediates of these coenzyme-enzyme complexes have been useful in establishing the mode of attachment of the coenzyme.

Another example of a prosthetic group, which illustrates the diversity of nonprotein components of the active center, is the functional pyruvate residue. Snell and his coworkers showed that a histidine decarboxylase, unlike many other decarboxylases, is not dependent on pyridoxal phosphate (Fig. 3.7, reaction b). Instead of pyridoxal phosphate, the functional aldehyde residue of a lactobacillus enzyme is provided by a covalently bound pyruvyl residue (28,29). It was shown that histidine decarboxylase of *Lactobacillus* 30 A is completely inhibited by reduction with $NaBH_4$ and by treatment with phenylhydrazine. When tritiated borohydride was used, labeled lactate was detected in the acid hydrolysate of the reduced enzyme. To establish that the pyruvate was the active aldehyde in analogy to pyridoxal phosphate, Recsei and Snell reduced the enzyme in the presence of substrate (histidine) and product (histamine) (30). Alkylated histidine and histamine derivatives were isolated from the acid hydrolysate, which proved the existence of the corresponding Schiff bases.

C. METAL IONS

Approximately 35% of the known enzymes contain tightly bound metal ions or are activated by metal ions. The enzymes having tightly bound metal ions, often at the active center, are termed *metallo-enzymes* (31). The common metal ions in metallo-enzymes are—Mn, Fe, Co, Ni, and Zn. Coordination of metal ions by metallo-enzymes can be extremely tight (K_d as low as 10^{-20}). Other enzymes form complexes with metal ions that are normal constituents of physiological tissues—for example, Na^+, K^+, Mg^{2+}, and Ca^{2+}. These *metal-enzyme complexes* are more labile than *metallo-enzymes* (K_d as high as 10^{-2}).

The study of Zn-metallo-enzymes has been pioneered and developed to a large extent by B. L. Vallee and his colleagues. Zinc is colorless and not detectable by EPR spectroscopy. The first zinc-containing enzyme was discovered somewhat fortuitously by Keilin and Mann in 1940 (32). These workers were studying a protein that they suspected contained copper because of its blue color. The analytical method they were using was adaptable to zinc detection with only minor changes in conditions. This fortunate circumstance led to the discovery of the first zinc-containing enzyme, carbonic anhydrase (the blue protein was hematocuprein). Since the discovery of carbonic anhydrase, at least 50 more zinc-containing enzymes have been discovered, including carboxypeptidases A and B, thermolysin, DNA polymerase, alcohol dehydrogenase, alkaline phosphatase, and aspartate transcarbamylase.

The detection and study of metal ions is not easy, mainly because of contamination and sensitivity problems. Inhibition by chelating agents such as 1,10-phenanthroline and EDTA usually provides the first clue that one or more metal ions is important. Atomic absorption spectrometry of protein samples free of exogenous metal is now commonly used to detect and quantitate metal constituents. The difficulty of metal detection can be illustrated by the recent discovery of Ni^{2+} in jack bean urease. In 1926, Sumner crystalized urease and, as a consequence of this discovery, suggested that enzymes were proteins and did not necessarily require metal ions (33). Nearly a half-century later Zerner and his co-workers discovered 2 g-atom of nickel per 105,000 g of enzyme (34).

Modern methods for the study of metal ions in enzymes include X-ray crystallography, optical and circular dichroism spectroscopy, electron paramagnetic resonance, and NMR spectroscopy (34a). The fact that manganese occurs at the active centers of many enzymes and can be replaced by Mg, Co, and so on has been exploited by Cohn, Mildvan, and others who have applied NMR techniques with great success. Consider the case of an enzyme-manganese-ligand complex. Changes in the relaxation rates of metal-bound water protons can be detected and used to study the active center. The nature of the ligand field of the bound metal ion and the accessibility of the metal ion to water have been elucidated. Also, the effect of the paramagnetic metal ion on protons of rapidly exchanging inhibitors, substrates, or coenzymes has been used to define crucial distances at the active center (34b).

Metal ions serve several functions in proteins. The metal ion coordinated at the active center can act as a general acid, which is vital in catalysis. On the basis of X-ray diffraction studies of an enzyme inhibitor complex, Lipscomb and coworkers proposed that the Zn^{2+} in carboxypeptidase A acts as a Lewis acid (35). In this case the zinc ion is coordinated to two histidine residues and a carboxylate group of glutamic acid. Vallee and Williams proposed that atypical coordination properties of Zn^{2+} and other metal ions in proteins reflect the "entatic state" (36). This term means that the metal ion is in a strained or tense state before interaction with substrate. In other words, inorganic ions can possess unusual reactivity as a result of their positions in biopolymers in a manner comparable to organic amino acid side chains. Mildvan and coworkers have proposed an essential Mg^{2+}-deoxynucleoside triphosphate complex. The magnesium ion is vital in aligning the 3' hydroxyl group of the growing DNA strand with the incoming nucleotide triphosphate (37).

D. CONCLUSIONS

Most enzyme active centers are complex. Binding sites are composed of subsites, which provide several points for enzyme-substrate interaction. Multipoint attachment is fundamental in explaining enzyme specificity. As was mentioned in Chapter 1, an apolar component near an ionic site is a frequent arrangement in substrate-binding sites. Arginine, lysine, aspartic acid, and glutamic acid are often found at the binding site near apolar pockets.

Several catalytic groups are usually important in enzymic catalytic sites. Consider chymotrypsin (asp-102, His-57, Ser-195), carboxypeptidase A (Tyr-248, Glu-270, and Zn^{2+}) and glyceraldehyde-3-phosphate dehydrogenase (Cys-149, His-176, and NAD) as examples of enzymes with three active-site constituents participating in the catalytic process. Invariably, two or more catalytic components are found at the active site.

Some amino acid side chains occur more frequently than others at the active center. The imidazole group of histidine is frequently present in the role of general base (pK 7). This residue can also act as a nucleophile in reactions such as phosphoryl transfer. The thiol group of cysteine is a powerful nucleophile and occurs at the active sites of thiol proteases and glyceraldehyde-3-phosphate dehydrogenase. The hydroxyl group of serine often occurs as an important nucleophilic group in enzyme active sites (serine proteinases, phosphatases, and phosphoglucomutase). The catalytic roles of these residues, coenzymes, and metal ions will be discussed in more detail in the following chapter and in Chapter 6.

REFERENCES

1. G. E. Means and R. E. Feeney, *Chemical Modification of Proteins,* Holden-Day, San Francisco (1971), p. 24.
2. A. M. Crestfield, W. H. Stein, and S. Moore, *J. Biol. Chem.* **238,** 2413 (1963).
3. A. M. Crestfield, W. H. Stein, and S. Moore, *J. Biol. Chem.* **238,** 2421 (1963).
4. A. W. Eyl and T. Inagami, *J. Biol. Chem.* **246,** 738 (1971).
5. R. M. Stroud, L. M. Kay, and R. E. Dickerson, *Cold Spring Harbor Symp. Quant. Biol.* **36,** 125 (1971).
6. B. S. Hartley and B. A. Kilby, *Biochem.* **56,** 288 (1954).

7. R. A. Oosterbaan, M. van Adrichem, and J. A. Cohen, *Biochem. Biophys. Acta* **63**, 204 (1964).
8. R. N. Perham and J. I. Harris, *J. Mol. Biol.* **7**, 316 (1963).
9. O. P. Molhatra and S. A. Bernhard, *J. Biol. Chem.* **243**, 1243 (1968).
10. B. R. Baker, W. W. Lee, W. S. Skinner, A. P. Martinez, and E. Tong, *J. Med. Pharm. Chem.* **2**, 633 (1960).
11. B. R. Baker, W. W. Lee, E. Tong, and L. O. Ross, *J. Am. Chem. Soc.* **83**, 3713 (1961).
11a. H. P. Meloche, *Biochemistry* **6**, 2273 (1967).
12. G. Schoellman, and E. Shaw, *Biochem. Biophys. Res. Commun.* **7**, 36 (1962).
13. G. Schoellman, and E. Shaw, *Biochemistry* **2**, 252, 1963.
14. F. J. Kezdy, A. Thomson, and M. L. Bender, *J. Am. Chem. Soc.* **89**, 1004 (1967).
15. E. B. Ong, E. Shaw, and G. Schoellman, *J. Am. Chem. Soc.* **86**, 1271 (1964).
16. A. Singh, E. R. Thornton, and F. H. Westheimer, *J. Biol. Chem.* **237**, 3006 (1962).
17. J. Schaefer, P. Baronowsky, R. Laursen, F. Finn, and F. H. Westheimer, *J. Biol. Chem.* **241**, 421 (1966).
18. C. S. Hexter, and F. H. Westheimer, *J. Biol. Chem.* **246**, 3928 (1971).
19. H. Bayley, and J. B. Knowles, *Methods Enzymol.* **46**, 69 (1977).
20. A. J. Bridges, and J. B. Knowles, *Biochem J.* **143**, 663 (1974).
21. N. Sonenberg, M. Wilchek, and A. Zamier, *Methods Enzymol.* **46**, 707 (1977).
22. R. R. Rando, *Methods Enzymol.* **46**, 28 (1977).
23. D. M. Blow, J. J. Birktoft, and B. S. Hartley, *Nature* **221**, 337 (1969).
24. D. M. Blow, *Enzymes* **3**, 185 (1970).
25. M. G. Rossman. A. Liljas, C. I. Branden, and L. J. Banaszak, *Enzymes* **11**, 61 (1975).
26. G. Biesecker, J. I. Harris, J. C. Thierry, J. E. Walker, and A. Wonacott, *Nature* **266**, 328 (1977).
27. D. Moras, K. W. Osen, M. N. Sabeson, G. C. Ford, and M. G. Rossman, *J. Biol. Chem.* **250**, 9137 (1975).
28. N. D. Riley and E. E. Snell, *Biochemistry* **7**, 3520 (1968).
29. N. D. Riley and E. E. Snell, *Biochemistry* **9**, 1485, (1970).
30. P. A. Recsei and E. E. Snell, *Biochemistry* **9**, 1491 (1970).
31. B. L. Vallee, *Adv. Prot. Chem.* **10**, 317 (1955).
32. D. Keilin and T. Mann, *Biochem J.* **34**, 1163 (1940).
33. J. B. Sumner, *J. Biol. Chem.* **69**, 435 (1926).
34. N. E. Nixon, C. Gazyola, R. L. Blakeley, and B. Zerner, *J. Am. Chem. Soc.* **97**, 4133 (1975).
34a. J. E. Coleman, in *Biochemistry Series One: Chemistry of Macromolecules* (H. Gutfreund, Ed.), Butterworth's, London (1974), p. 185.
34b. A. S. Mildvan and M. Cohn, *Adv. Enzymol.* **33**, 1 (1970).
35. F. A. Quiocho and W. N. Lipscomb, *Adv. Prot. Chem.* **25**, 1 (1971).
36. B. L. Vallee and R. J. P. Williams, *Proc. Natl. Acad. Sci. USA,* **59**, 498 (1968).
37. D. L. Sloan, L. A. Loeb, A. S. Mildvan, and R. J. Feldman, *J. Biol. Chem.* **250**, 8913 (1975).

CHAPTER FOUR

MECHANISMS OF ENZYME-CATALYZED REACTIONS

In this chapter we will look at the specific catalytic roles of amino acid side chains and other components of enzyme active centers. Three basic chemical mechanisms—general acid-base, nucleophilic, and electrophilic catalysis—will be discussed. Acid-base catalysis implies proton transfer to or from the reactant with a resultant acceleration of the reaction rate. Nucleophilic catalysis involves the nucleophilic attack of the catalyst on the reactant, with the formation of a covalent intermediate. An electrophilic catalyst accepts an electron pair from the substrate. As we shall see, a given enzymatic reaction may include all of these mechanisms. Much of what we know about these catalytic mechanisms comes from the study of simple models, which are organic compounds of low molecular weight. Detailed discussions of such studies may be found in works of Bruice and Benkovic (1), Jencks (2), and Bender (3).

The general concepts of approximation, alignment, and destabilization are also discussed in this chapter.

A. GENERAL ACID-BASE CATALYSIS

Organic acids and bases can catalyze chemical reactions by donation or acceptance of protons. Protons and hydroxide ions can also act catalytically, as follows:

$$A + H^+ \rightleftharpoons AH \rightarrow \text{Product} + H^+$$
$$A + OH^- \rightleftharpoons AOH \rightarrow \text{Product} + OH^-$$

This catalysis by H^+ or OH^- is called *specific acid* or *specific base catalysis*. When a reaction is catalyzed solely by protons or hydroxide ions, the rate is not dependent on buffer concentration. Jencks and Carriuolo studied the hydrolysis of acetylimidazole in the presence of imidazole buffer:

$$CH_3C(O)-\text{Im} \xrightarrow{H_2O, \text{HN-Im}} CH_3CO_2H + \text{imidazole}$$

They found the rate, at a constant pH, to be dependent upon buffer concentration (Fig. 4.1). The dependence of rate on buffer concentration cannot be related to specific acid or specific base catalysis because $[H^+]$ and $[OH^-]$ are constant

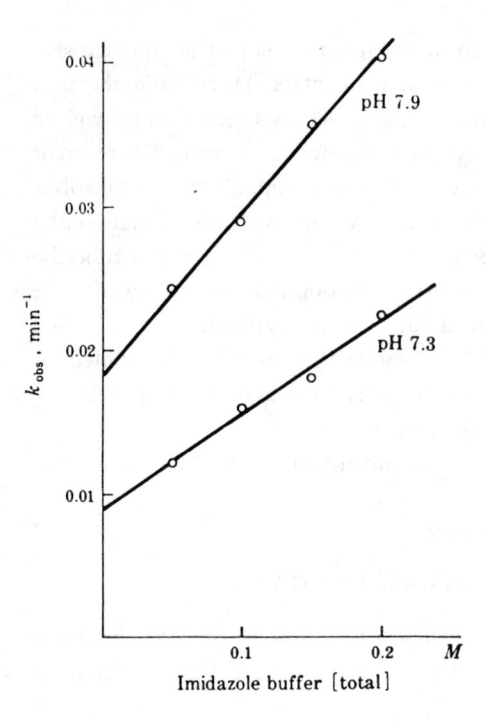

Figure 4.1. Dependence of the rate of acetylimidazole hydrolysis as a function of buffer concentration (2).

A. GENERAL ACID-BASE CATALYSIS

at a given pH, regardless of the concentration of the buffer components. Could the buffer effect result from imidazole's acting as a nucleophile? This possibility is ruled out because, were that the case, the product, acetylimidazole, would be identical to the substrate.

The explanation of the data shown in Figure 4.1 is that the basic component of the buffer acts as a general base catalyst in the following way:

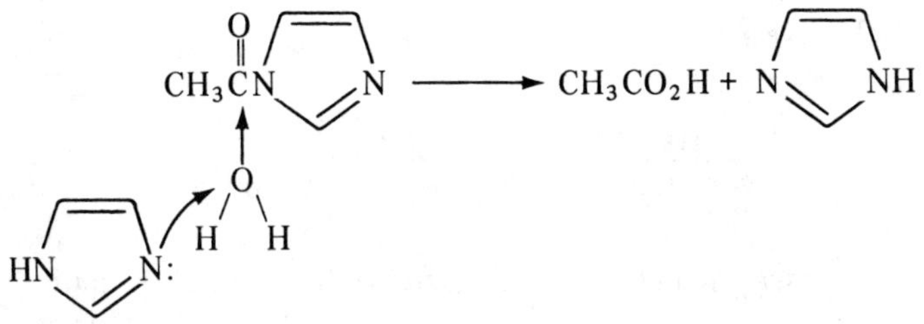

$$CH_3CN\text{-imidazole} \longrightarrow CH_3CO_2H + \text{imidazole}$$

At a given pH an increase in buffer concentration results in an increase in the concentration of imidazole, which leads to a faster rate. It is apparent that imidazole (free base) is the active catalytic species; the rate dependence on buffer concentration is higher at pH 7.9 than at pH 7.3. Notice (Figure 4.1) that the rate at zero buffer concentration is higher at pH 7.9 than at pH 7.3. This means that specific base catalysis is also a factor in this particular reaction. The fact that the reaction goes on in the absence of buffer at pH 7 means that there is a rate term for the solvent (water) itself. Could the proton and acidic buffer species, imidazolium ion, also be involved? In some cases the answer is yes. The rate equation for such a system would be:

$$\text{rate} = k_{solv}[S] + k_{H^+}[H^+][S] + k_{OH^-}[S] + k_{BH^+}[BH^+][S] + k_B[B][S]$$

$$= (k_0 + k_b)[S]$$

$$= k_{obs}[S]$$

Therefore, k_0 is the rate constant in the absence of buffer

$$k_0 = k_{solv} + k_{H^+} \cdot [H^+] + k_{OH^-} \cdot [OH^-]$$

The buffer terms constitute k_b.

$$k_b = k_{BH^+} \cdot [BH^+] + k_B \cdot [B]$$

When the buffer concentration is equal to zero, $k_{obs} = k_0$. Extrapolation to zero buffer concentration (Fig. 4.1) permits the determination of k_0. When k_0 is known, k_B can be calculated:

$$(k_{obs} - k_0) = k_b = k_{BH^+} \cdot [BH^+] + k_B [B]$$

Division of the above equation by B_T (total concentration of buffer species, $([B] + [BH^+])$, gives:

$$\frac{(k_{obs} - k_0)}{B_T} = k_{BH^+} \frac{[BH^+]}{B_T} + k_B \cdot \frac{[B]}{B_T}$$

The rate constants, k_{BH^+} and k_B, can be found with the aid of the plot

$$(k_{obs} - k_0)/B_T \text{ vs } [BH^+]/B_T \text{ or } (k_{obs} - k_0)/B_T \text{ vs } [B]/B_T$$

(Fig. 4.2). When $[B]/B_T$ equals one (all the buffer is the free base), $[BH^+]/B_T$

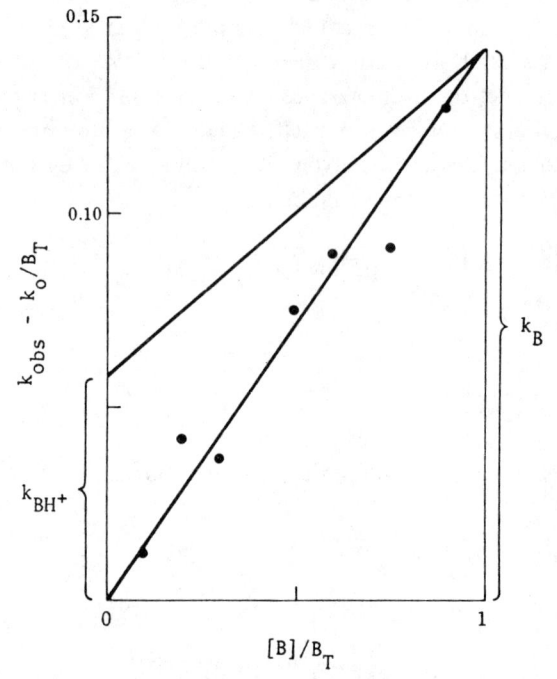

Figure 4.2. Plot of $(k_{obs} - k_0)/B_T$ versus $[B]/B_T$. B_T is the total buffer concentration (2).

A. GENERAL ACID-BASE CATALYSIS

equals zero and the value of $(k_{obs} - k_0)/B_T$ equals k_B. When $[B]/B_T$ equals zero, $[BH^+]/B_T$ equals one and $(k_{obs} - k_0)/B_T$ equals k_{BH^+}. The lower line in Figure 4.2 intersects zero, which means k_{BH^+} equals zero or that the acidic species of the buffer does not contribute to catalysis. This circumstance holds for the imidazole-catalyzed hydrolysis of acetylimidazole, in which case only general-base catalysis occurs.

General-acid catalysis can occur in concert with general-base catalysis in the following way:

$$B: \overset{H}{\underset{H}{\diagdown}} O \cdots \overset{R}{\underset{X}{C}}=O \cdots H-B^+ \longrightarrow HO\overset{R}{\underset{X}{C}}OH + B$$

$$\downarrow$$

$$RCO_2H + HX$$

Note that in this concerted mechanism that catalytic species are not altered during the course of the reaction. This is an important point when rates of proton transfer are considered. The donation of a proton in a single-step, general acid-catalyzed reaction must be followed by reprotonation of the general acid. If the reprotonation by water is relatively slow, the contribution of the general-acid step would be of no advantage. Given below are several proton transfer reactions, along with forward and reverse rates:

$$OH^- + H-N\diagup\diagdown^+N-H \underset{k_{rev} = 1.2 \times 10^3 sec^{-1}}{\overset{k_{for} = 2.5 \times 10^3 sec^{-1}}{\rightleftharpoons}} H_2O + N\diagup\diagdown NH$$

$$OH^- + CH_3NH_3^+ \underset{k_{rev} = 2 \times 10^4 sec^{-1}}{\overset{k_{for} = 4 \times 10^3 sec^{-1}}{\rightleftharpoons}} H_2O + CH_3NH_2$$

$$OH^- + \langle\bigcirc\rangle-OH \underset{k_{rev} = 2 \times 10^3 sec^{-1}}{\overset{k_{for} = 1.4 \times 10^3 sec^{-1}}{\rightleftharpoons}} H_2O + \langle\bigcirc\rangle-O^-$$

$$HN\diagup\diagdown N + H_3O^+ \underset{k_{rev} = 0.8 \times 10^3 sec^{-1}}{\overset{k_{for} = 1.5 \times 10^3 sec^{-1}}{\rightleftharpoons}} H-N\diagup\diagdown^+N-H + H_2O$$

The forward and reverse rates of these reactions are of the same order of magnitude. These examples would suggest that 10^3 sec^{-1} is a characteristic value. It is significant that many enzyme-catalyzed reactions that contain proton transfer steps have turnover numbers near this value of 10^3 sec^{-1}, which suggests that proton transfer may be rate-limiting.

B. NUCLEOPHILIC CATALYSIS

Nucleophilic catalysis involves the donation of an electron pair with the formation of a covalent intermediate. In the late 1950s and early 1960s imidazole-catalyzed acyl transfer was studied because of the presence of histidine at the active site of α-chymotrypsin (5-7). As it happens, the imidazole ring does not function as a nucleophile in the chymotrypsin-catalyzed acyl transfer, but rather as a general base. However, histidine does function as a nucleophile in other enzymatic reactions, and the model studies with activated acyl compounds have been of considerable value.

Active acyl compounds such as nitrophenylacetate are subject to nucleophilic catalysis by imidazole. The reaction proceeds with the formation of an acyl imidazole intermediate, which can transfer the acyl group to water or to some other nucleophile:

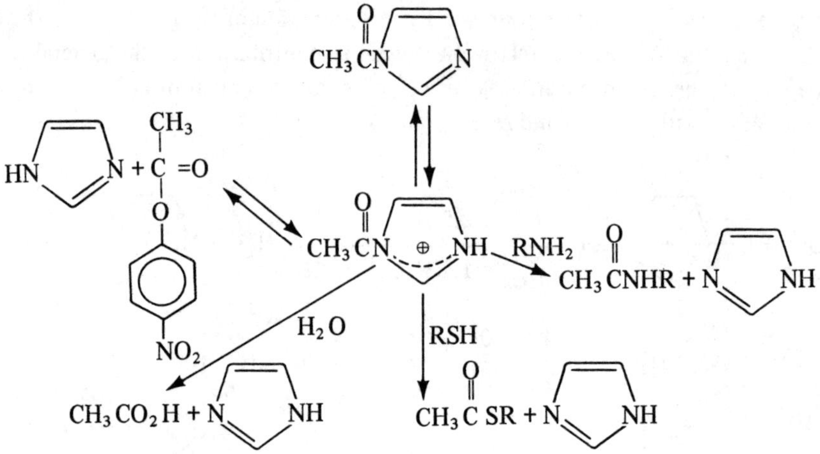

The acyl intermediate has an absorption maximum of 245nm and can thus be verified experimentally. The reactive intermediate is the protonated acyl imidazolium ion.

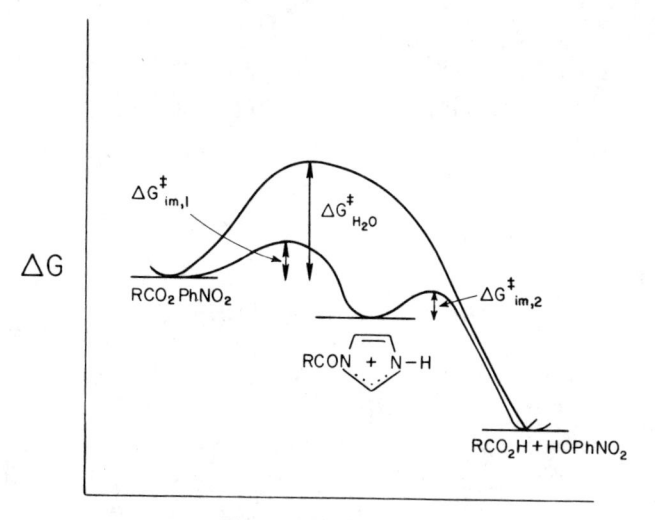

B. NUCLEOPHILIC CATALYSIS

Figure 4.3. Free energy diagram for the hydrolysis of an ester in the presence (bottom curve) and absence of imidazole (2).

A free energy diagram of this reaction is shown in Figure 4.3. There are a number of important points illustrated in this diagram.

1. *The catalyst must have greater nucleophilic reactivity than the final acyl acceptor.* If the rate of reaction of water with the substrate were faster than the rate of imidazole attack, there would be no catalysis. The activation free energy, $\Delta G_{im,1}^{\ddagger}$ must be less than that for the water reaction, $\Delta G_{H_2O}^{\ddagger}$, in Figure 4.3.

2. *The covalent intermediate must be more reactive than the substrate.* For catalysis to occur, intermediates along the catalytic reaction path must react faster than the substrate in the uncatalyzed reaction, $\Delta G_{im,2}^{\ddagger} < \Delta G_{H_2O}^{\ddagger}$.

3. *The covalent intermediate must be thermodynamically less stable than the products.* Regardless of rates, the free energy level of the covalent intermediate must be above that of the product. Otherwise the intermediate would accumulate and the reaction would not result in the formation of the desired product.

Other examples of nucleophilic catalysis by molecules of low molecular weight follow:

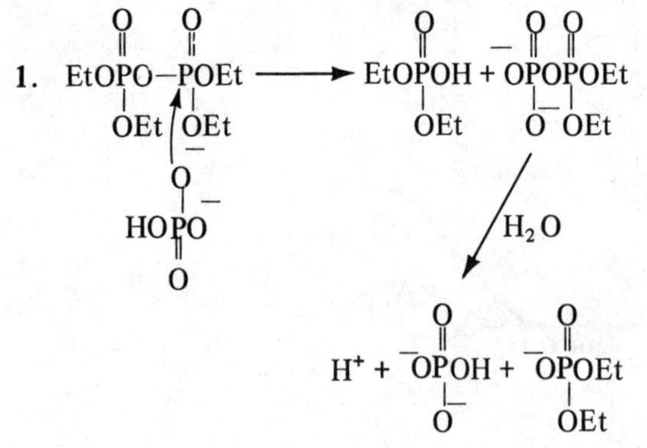

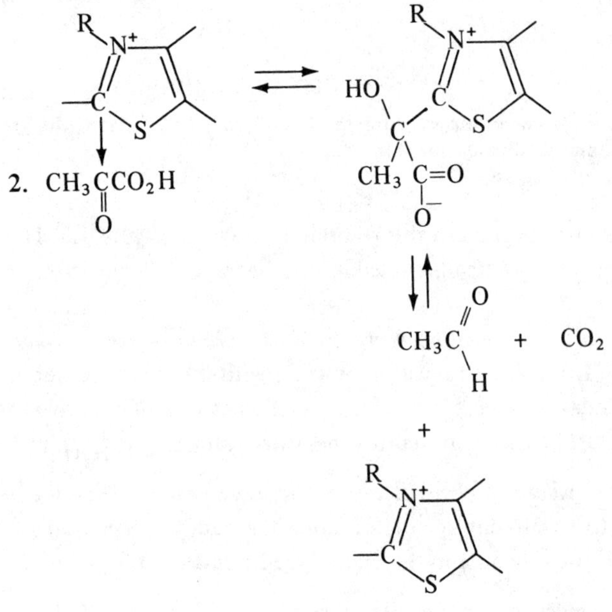

1.
$$\text{EtOPO–POEt} \longrightarrow \text{EtOPOH} + \text{OPOPOEt}$$
(with OEt groups)

$$\text{HOPO} \quad / \quad H_2O$$

$$H^+ + \text{}^-\text{OPOH} + \text{}^-\text{OPOEt}$$

2.

$$\text{CH}_3\text{CCO}_2\text{H} \rightleftharpoons \text{(HO–C intermediate)} \quad \text{CH}_3 \, C=O$$

$$\updownarrow$$

$$\text{CH}_3\text{C(=O)H} + CO_2$$

$$+$$

(thiazolium ylide)

Amino acid side chains that can act as nucleophiles include the following: cysteine thiol, serine hydroxyl, ε-amino group of lysine, imidazole ring of histidine, carboxyl groups of aspartic and glutamic acid, and the phenolic hydroxyl group of tyrosine. It is obvious that groups such as these can also act as general bases or general acids. A problem arises in distinguishing the nucleophilic mechanism

B. NUCLEOPHILIC CATALYSIS

3. 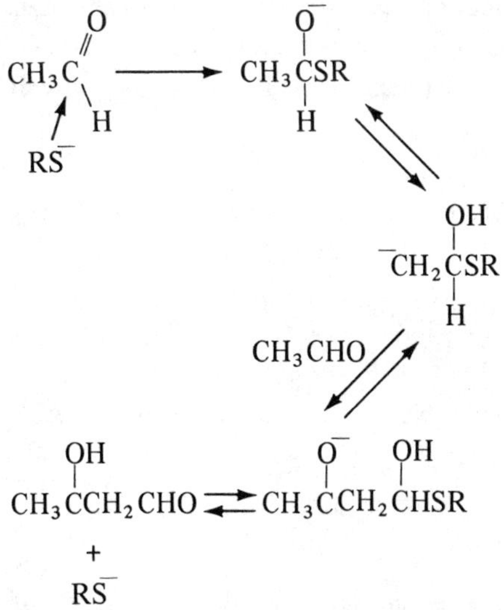
$$CH_3\overset{O}{\underset{\underset{RS^-}{\uparrow H}}{C}} \longrightarrow CH_3\overset{O^-}{\underset{H}{\underset{|}{C}SR}} \rightleftarrows \,^-CH_2\overset{OH}{\underset{H}{\underset{|}{C}SR}}$$

$$CH_3CHO \nearrow$$

$$\underset{+\,RS^-}{CH_3\overset{OH}{\underset{|}{C}}CH_2CHO} \rightleftarrows CH_3\overset{O^-}{\underset{|}{C}}CH_2\overset{OH}{\underset{|}{C}}HSR$$

from the general base mechanism. In reactions with small molecules we have already seen that catalysis by the leaving group requires that the mechanisms be general-base, rather than nucleophilic catalysis:

$$RCB \xrightarrow[H_2O]{B} RCO_2H + HB$$

Nucleophilic attack by B would result in no change of the reactant. The specific example cited earlier was imidazole attacking acetylimidazole. The detection of a covalent intermediate indicates nucleophilic catalysis. Solvent deuterium isotope effects can be of value in supporting one mechanism over the other. In the acyl transfer reaction shown below, the presence of D_2O will slow the reaction rate:

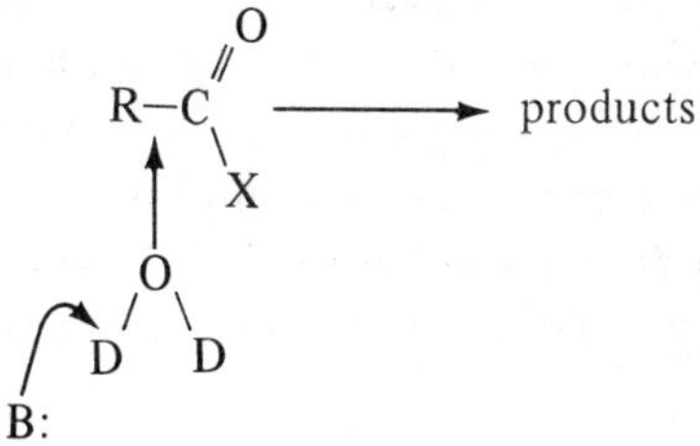

$$R-C\overset{O}{\underset{X}{\Vert}} \longrightarrow products$$

B:

Transfer of the heavier deuteron is slower than the transfer of a proton. When the proton transfer step is rate-determining, the value of k_{H_2O}/k_{D_2O} will be 2-3. In nucleophilic catalysis, the value of k_{H_2O}/k_{D_2O} will be 1. Examples (8-10) follow.

$$RC(=O)OphNO_2 \xrightarrow{\text{HN} \diagup \diagdown \text{N}} \text{products}; \quad k_{H_2O}/k_{D_2O} = 1$$

$$CHCl_2C(=O)OEt \xrightarrow{\text{HN} \diagup \diagdown \text{N}} \text{products}; \quad k_{H_2O}/k_{D_2O} = 3$$

Bender and coworkers showed that the deacylation of α-chymotrypsin went faster in H_2O than in D_2O (c.f. reference 11). Reactions with a number of substrates, including specific substrates, exhibited solvent deuterium isotope effects in the range of 2 to 3. In the case of α-chymotrypsin the results of solvent-deuterium isotope studies were quite valuable, and indicated that His-57 acted as a general base rather than as a nucleophile. Unfortunately, isotope effects are not always straightforward. D_2O perturbation of pK_a's, protein structure, and substrate binding are potential problems to be considered.

The effectiveness of general-base catalysts and nucleophilic catalysts is dependent upon basicity, at least for the ester hydrolyses shown above. This dependence may be expressed by the Brønsted equation

$$\log k/k_0 = \beta pK + C$$

The term k_0 refers to the rate constant of the reaction of a reference compound. For a series of general bases or nucleophiles a plot of $\log k/k_0$ versus pK should give a straight line with a slope of β. C is a constant characteristic of the reaction. For nucleophilic reactions a value of β near one indicates that basicity is a good model for nucleophilic reactivity and that the nucleophile would acquire a positive charge in the transition state. For the hydrolysis of nitrophenyl acetate the Brønsted β is 0.8 (Fig. 4.4), which suggests that the transition state looks like this:

$$\text{Nuc}^+ \cdots\cdots\cdots \overset{\overset{O}{\|}}{C} \cdots\cdots\cdots {}^-OpHNO_2$$

B. NUCLEOPHILIC CATALYSIS

For nucleophilic catalysis, deviations from the line of the Brønsted plot are large compared to those generally observed for general-base catalysis. In Figure 4.4, for example, the rate constant for Tris (trishydroxymethyl aminomethane) falls below the line. Tris is a hindered amine and so reacts more slowly than one might predict from the pK_a. When acting as a proton acceptor, hindered bases behave in a predictable way because protonation is not retarded. It is known that nitrogen nucleophiles and oxyanions differ considerably in their nucleophil-

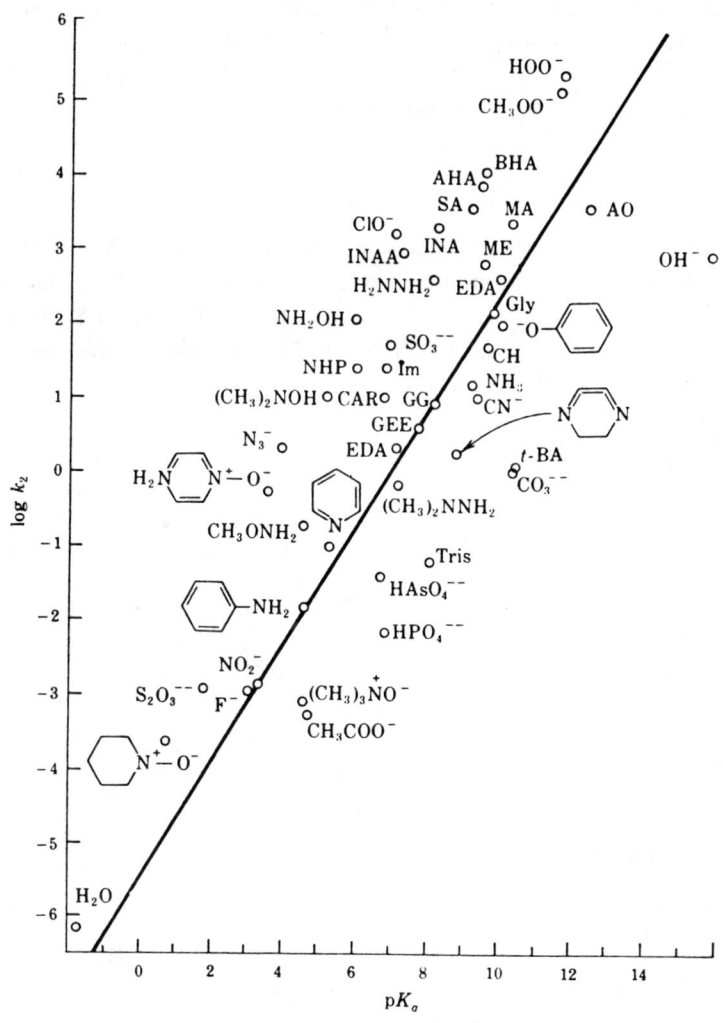

Figure 4.4. The cleavage of nitrophenyl acetate by a variety of nucleophiles (2).

icity. Pyridine, for example, has a pK_a similar to that of acetate. As is shown in Figure 4.4, the rate constant for pyridine falls on the line, but that for acetate anion is low by two powers of ten. Positive deviations from the line may be explained by the "α-effect." In many cases a compound such as an oxime (–C=N–O⁻) is an unusually potent nucleophile because the atom next to the nucleophilic atom possesses a free electron pair. In Figure 4.4 hydroxylamine, hydrazine, and peroxy anions are examples of α-effect compounds. Such effects are important in catalysis by enzymes and synthetic polymers. In these cases the enhanced nucleophilicity can result from assistance by free electron pairs positioned near the nucleophilic atom as a result of the three-dimensional conformation of the polymer. Groups widely separated in the linear sequence of a protein may be juxtaposed by the folding of the peptide backbone.

C. ELECTROPHILIC CATALYSIS

"Electron sink" effects are abundant in enzymatic catalysis. Pyridoxal phosphate functions as an electrophile in several reaction types (Chapter 3), of which decarboxylation of an amino acid is an example. In this reaction a Schiff base is formed between the aldehyde and the amine. The pyridine ring then serves as an electron sink for the subsequent reaction

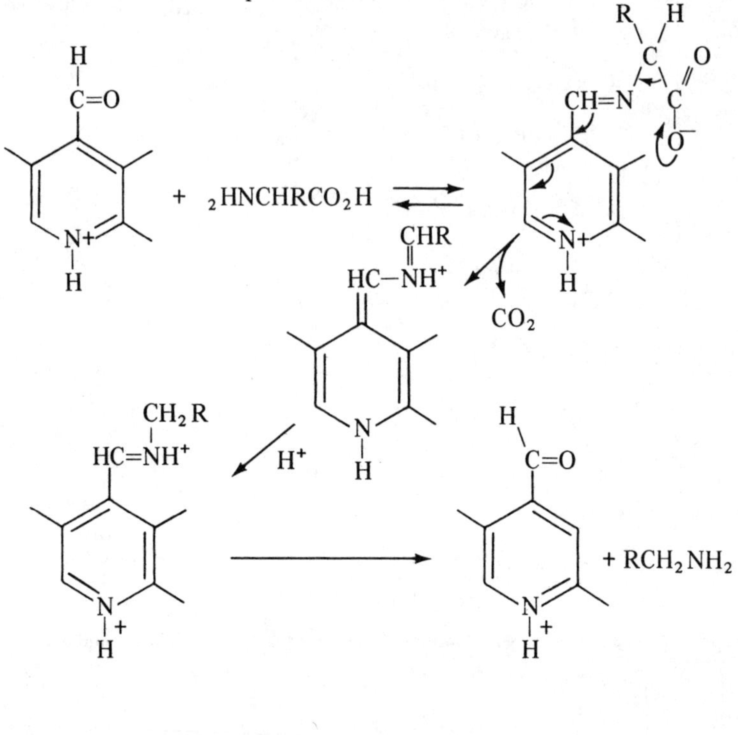

C. ELECTROPHILIC CATALYSIS

Amino groups at the active center can form Schiff bases with carbonyl substrates. Aldolase, δ-aminolevulinate dehydratase, and acetoacetate decarboxylase are enzymes for which this mechanism has been proposed. The case of acetoacetate decarboxylase is illustrated here:

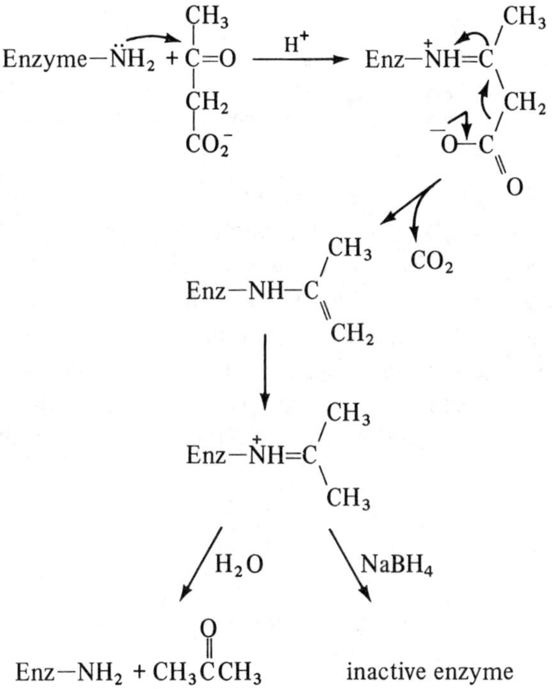

$$\text{Enzyme-NH}_2 + \underset{\underset{CO_2^-}{\overset{CH_2}{|}}}{\overset{CH_3}{\underset{|}{C=O}}} \xrightarrow{H^+} \text{Enz-NH}^+=C\overset{CH_3}{\underset{CH_2}{\diagdown}}{\overset{\diagup}{}} \cdots O-C=O$$

$$\text{Enz-NH-C}\overset{CH_3}{\underset{CH_2}{\diagdown}} \quad CO_2$$

$$\downarrow$$

$$\text{Enz-}\overset{+}{N}H=C\overset{CH_3}{\underset{CH_3}{\diagdown}}$$

$$\swarrow H_2O \qquad \searrow NaBH_4$$

$$\text{Enz-NH}_2 + CH_3\overset{\overset{O}{\|}}{C}CH_3 \qquad \text{inactive enzyme}$$

The protonated Schiff base is readily reduced by sodium borohydride, which results in an isopropyl derivative of lysine (12). This adduct is stable on hydrolysis of the protein with 6N HCl, and can therefore be identified.

Metal ions (Zn^{2+}, Mn^{2+}, Co^{2+}, etc.) can serve as electron sinks at enzyme active centers. Using magnetic resonance methods, Mildvan and coworkers have proposed the model shown in Figure 4.5 for the polarization of the carbonyl group in metallo-enzymes (13). Pyruvate carboxylase, transcarboxylase, malic enzyme, alcohol dehydrogenase, ribulose diphosphate carboxylase, and mandelate race-

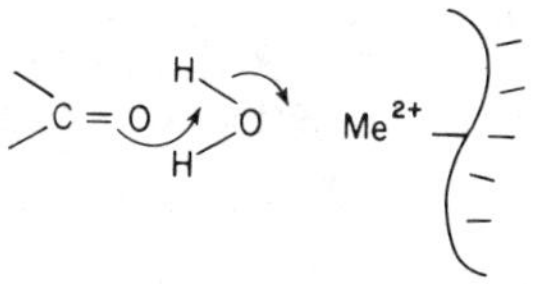

Figure 4.5. Polarization of the carbonyl group by enzyme-bound metal ion.

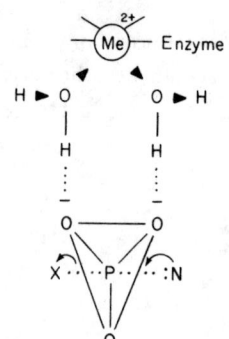

Figure 4.6. Nucleophilic reaction of phosphoryl phosphorus as influenced by water bound to a metal ion in an enzyme active center.

mase are believed to have such an arrangement of carboxyl group–water–metal ion. Second-sphere complexes have also appeared in phosphoryl- and nucleotidyl-transfer enzymes. Based on the results with DNA polymerase (I), pyruvate kinase, phosphoglucomutase, and the Na^+,K^+-ATPase, the model shown in Figure 4.6 was suggested by Mildvan and coworkers (14). In thise case two metal-bound water molecules activate the phosphoryl group for nucleophilic attack by N.

The zinc ion at the active center of carboxypeptase A was thought to act as an electrophile, as shown in Figure 4.7 (15). Direct coordination of the carbonyl

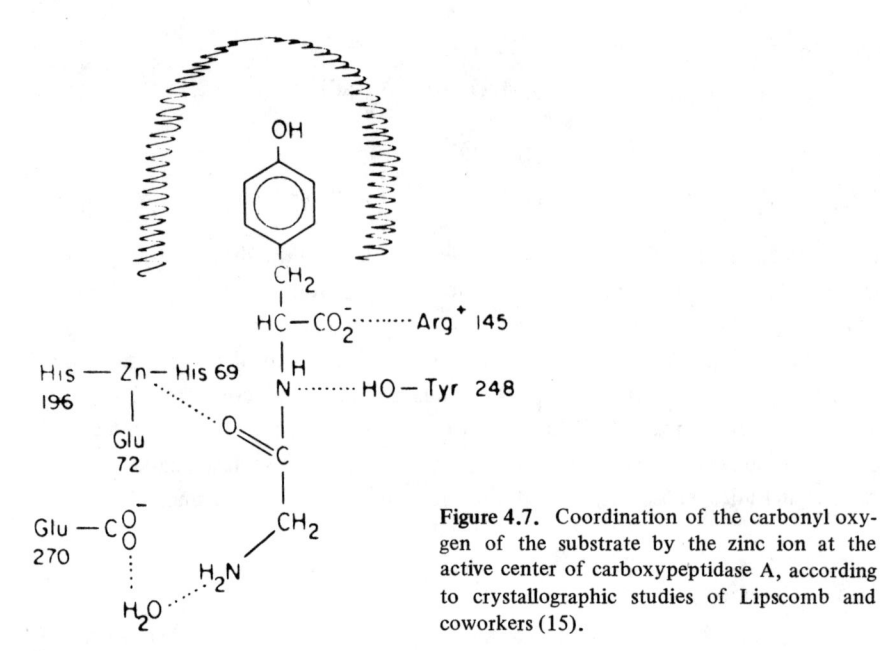

Figure 4.7. Coordination of the carbonyl oxygen of the substrate by the zinc ion at the active center of carboxypeptidase A, according to crystallographic studies of Lipscomb and coworkers (15).

oxygen was believed to activate the amide group for nucleophilic attack by either the carboxyl group of Glu-270 or by water as assisted by Glu-270 acting as a general base. As we shall see in the next section, there is considerable recent evidence that the Zn^{2+} atom acts to coordinate water, which attacks an anhydride intermediate formed between the acyl compound of the substrate and the carboxyl group of Glu-270. This idea is supported by model studies that have shown formidable rate enhancements brought about by metal-bound hydroxide.

D. EXAMPLES OF SPECIFIC ENZYME MECHANISMS

Serine Proteases

Serine proteases have been actively studied in terms of mechanism for many years. Through the efforts of crystallographers, kineticists, spectroscopists, and others, much has been learned about this class of enzymes. The most popular of this class are α-chymotrypsin and trypsin. These enzymes are very similar structurally, with the exception that there is a carboxyl group at the bottom of the hydrophobic binding pocket in trypsin. Both enzymes have essential histidine and serine residues. Another characteristic of the serine proteases is a catalytic triad composed of an aspartate residue, imidazole ring of histidine, and the hydroxymethyl group of serine, shown in Figure 4.8. This characteristic arrangement has been suggested for other serine proteases from widely different sources, including microbial species (16-20). The Blow group suggested the "charge relay" system, whereby a proton from serine goes to the imidazole ring and from there to the carboxyl group of Asp-102 (21). The alkoxide ion produced by this scheme would then attack the substrate to give a tetrahedral intermediate, which would then go to the acyl enzyme. In the "charge relay" model the carboxylate ion would take a proton from an imidazolium ion, which would not happen in solution because imidazole is a stronger base than a carboxylate. Using ^{13}C NMR, Richards and collaborators collected data that led them to suggest that the pK_a's of the imidazole and carboxyl group were reversed as a result of a special microenvironment at the enzyme active center (22). Additional NMR spectroscopy

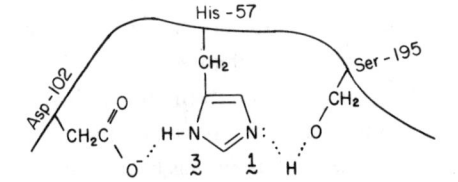

Figure 4.8. The catalytic triad of serine proteases, showing the carboxylate ion of Asp-102 stabilizing the reactive tautomer of the imidazole of His-57.

studies by Markley and Porubcan (23) and by Robillard and Shulman (24) created controversy. Infrared studies supported the idea of a reversal of ionizations (25).

Bachovchin and Roberts studied ^{15}N-NMR spectroscopy of an α-lytic protease; the single histidine of this enzyme was enriched in ^{15}N in the imidazole ring (26). The chemical shifts (^{15}N) indicated that the pK_a of the histidyl residue was 7.0 ± 0.1 at 26°C. These authors also concluded that the tautomer with the hydrogen on N3 of the imidazole ring predominates in the pH range where the enzyme is active. The hypothesis of charge transfer was not supported (26), although the general geometry of the catalytic triad (Fig. 4.8) is probably correct.

A summary of the mechanism of serine proteases would not be changed drastically by the modification of the role of Asp-102 discussed above. The imidazole of His-57 acts as a general base to abstract a proton from the hydroxyl group of Ser-195. The carboxylate of Asp-102 helps to align the imidazole ring and to stabilize the imidazolium cation that is formed. The alkoxide then attacks the amide carbon with the formation of the acyl enzyme, which probably involves a tetrahedral intermediate. The imidazole is deprotonated at this stage, having donated its proton to the leaving group. The deacylation step involves the attack of water, assisted by the neutral imidazole acting as a general base. After departure of the acyl group, the proton abstracted from water is shared between the imidazole ring and the serine hydroxyl group. This arrangement is the starting configuration of the active center shown in Figure 4.8.

Carboxypeptidase A

This is another amidase that has been studied very actively by scientists, using a variety of approaches and tools. The picture of the active center of the enzyme derived from crystallographic studies by Lipscomb and his collaborators appears in Figure 4.7 (15). Carboxypeptidase A is specific for peptides with a free carboxyl terminus. Peptides terminating in apolar residues are preferred substrates. The binding of the free carboxylate to Arg-145 and the side chain in an apolar pocket is illustrated in Figure 4.7. The roles of Tyr-248, Glu-270, and the zinc ion have been disputed, and even the equivalence of the crystal structure and the solution structure has been questioned (27). Vallee and coworkers found that the color of arsanilloazocarboxypeptidase was yellow in the crystalline state, but red in solution. The crystallographic studies reveal a large change in the position of Try-248 when the inhibitor, glycyl-tyrosine, binds. The conformational change presumably results in Tyr-248's being positioned close to the amide nitrogen, where it acts as a general acid (Fig. 4.7). The role of Glu-270 has been a matter

D. EXAMPLES SPECIFIC ENZYME MECHANISMS

of conjecture. The possibilities are direct nucleophilic attack to give an (anhydride) acyl intermediate, or general-base catalysis with an intervening water molecule. Both mechanisms are known in model compounds. Bruice and Pandit, for example, have documented the nucleophilic mechanism in simple organic acids [Table 4.1; (28)]. Fehrst and Kirby have shown that the hydrolysis of aspirin involves the carboxylate group's acting as a general base:

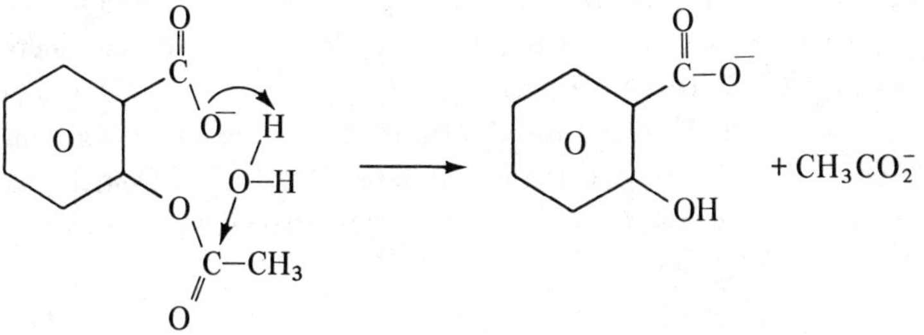

TABLE 4.1. Rates of Ester Hydrolysis and the Demonstration of Anchimeric Assistance

Ester	$\dfrac{k_{hydrol}}{k_{hydrol} \, (glutarate)}$
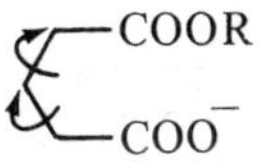 —COOR / —COO⁻	1
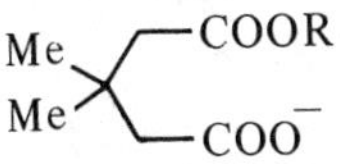 Me, Me —COOR / —COO⁻	20
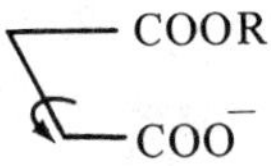 —COOR / —COO⁻	230
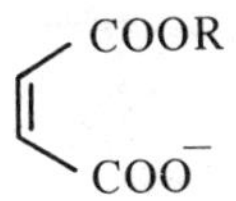 —COOR / —COO⁻	10,000
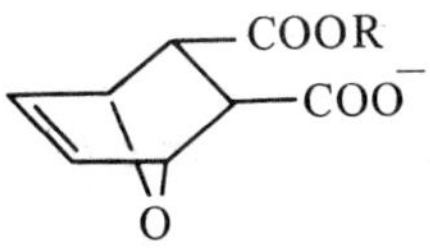 —COOR / —COO⁻	53,000

Solvent deuterium isotope effects would be expected if Glu-270 were to act as a general base, but such effects were not observed by Fife and Squillacote (30). However, using a specific ester substrate, Makinen and collaborators found a solvent deuterium isotope effect of 1.9 in k_{cat} (31). This work involved kinetic studies of the hydrolysis of O-(trans-p-chlorocinnamoyl)-L-β-phenyllactate and spectrophotometric studies at subzero temperatures. This study and earlier work (32) suggest a covalent intermediate. Moreover, the Makinen group suggests that the zinc ion acts as the hydroxide to promote hydrolysis of the mixed anhydride, which involves Glu-270. This mechanism appears in Figure 4.9. For a peptide substrate, however, Breslow and Wernick suggested that no covalent intermediate exists (33), because the enzyme did not catalyze ^{18}O exchange from the carboxyl component to the solvent. However, it is possible that the water molecule is re-

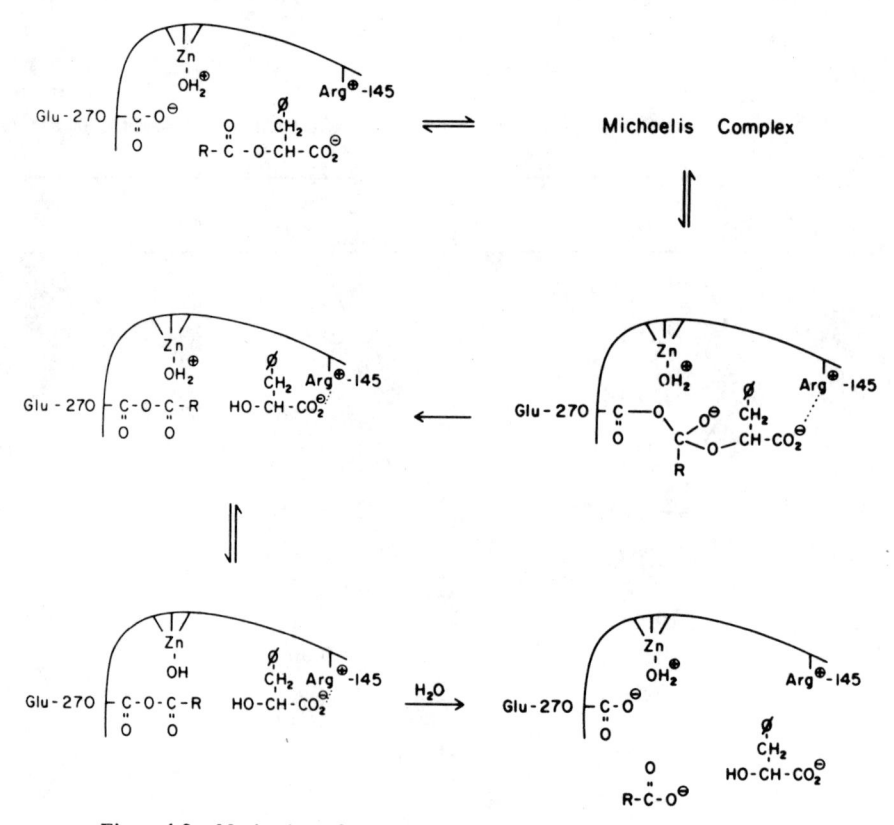

Figure 4.9. Mechanism of carboxypeptidase A, according to Makinen (31).

D. EXAMPLES SPECIFIC ENZYME MECHANISMS

tained by a specific site on the enzyme. Makinen and collaborators suggest that this site is the metal ion (31).

The promotion of hydrolytic reactions by metal ions has been shown in a number of metal-ion complexes of low molecular weight (34-36). An interesting example is shown below (36):

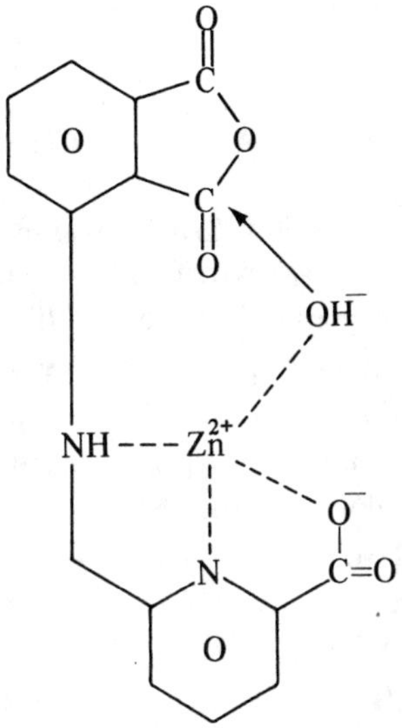

The hydrolysis of this anhydride proceeds 10^3 times faster in the presence of metal ion than in its absence.

The Makinen group observed a biphasic hydrolysis of the specific substrate 0-(trans-p-chlorocinnamoyl)-L-β-phenyllactate (31) at subzero temperatures. The rate constant, k_s, for the slow step extrapolates to k_{cat} in Arrhenius plots. This common slope of the Arrhenius plots suggests that the presence of the cryosolvent does not perturb the mechanism of the rate-limiting catalytic step. The pH dependence of k_s shows a metal ion-dependent pK_a—about pH 7.3 for the zinc enzyme and about pH 6.2 for the cobalt enzyme. The enthalpy change for the ionization was also metal-dependent: ΔH_{ion} equals 7.2 ± 0.1 kcal/mol for the zinc enzyme, and 6.2 ± 0.1 for the cobalt enzyme. This evidence suggests that the ionization is that of a metal-bound water molecule. The conclusion is that for hydrolysis of a specific ester substrate the deacylation of a mixed anhydride intermediate is promoted by a Zn^{2+} hydroxide (Fig. 4.8).

Lactate Dehydrogenase

The oxidation of L-lactate to pyruvate is catalyzed by lactate dehydrogenase, which is found in skeletal and heart muscle:

$$\begin{array}{c} CO_2H \\ | \\ HO-C-H \\ | \\ CH_3 \end{array} + NAD^+ \rightleftharpoons \begin{array}{c} CO_2H \\ | \\ C=O \\ | \\ CH_3 \end{array} + NADH + H^+$$

L (S)-lactate   pyruvate

The heart (H_4) and skeletal muscle enzymes (M_4) are tetramers and are termed isozymes (37), since they are two structurally distinguishable forms with the same enzyme activity. The kinetic parameters of the two enzyme forms differ. Despite the kinetic and structural differences, they will form hybrids with predictable statistical distribution of the various forms—M_3H_1, M_2H_2, M_1H_3. Rossman and collaborators have done crystallographic studies with the enzyme (M_4) from the skeletal muscle of dogfish (39).

The pH dependence of substrate binding (40) and chemical modification studies (41) strongly suggest an essential histidine residue at the active center. The active-site peptide containing alkylated histidine was sequenced by Woenckhaus and collaborators (42). The histidine of importance was found to be at position 195, a finding supported by the X-ray structure.

The enzyme substrate complex may be represented schematically by Figure 4.10 (39). The nicotinamide ring interacts with the hydrophobic side chains of Val-32 and Val-247. The stereochemistry of the reaction requires that the rotation around the glycosidic bond between the nicotinamide-ring nitrogen and the ribose be frozen. The carboxamido group provides a handle for freezing out this rotational mode; it is hydrogen bonded to the ammonium group of Lys-250.

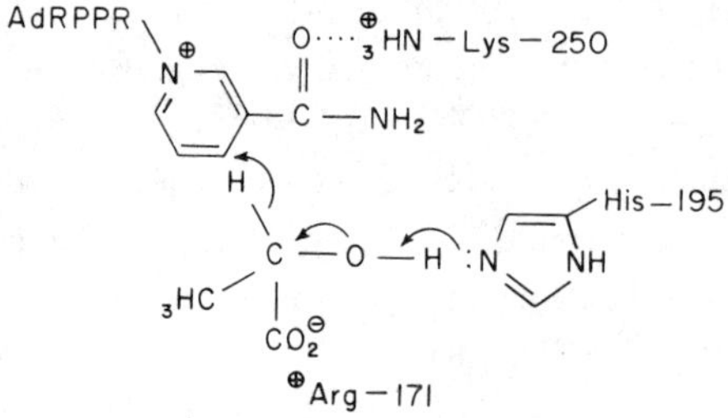

Figure 4.10. The productive ternary complex of NAD-lactate-lactate dehydrogenase.

D. EXAMPLES SPECIFIC ENZYME MECHANISMS

Modification of the nicotinamide ring at the 3 positions shows that the carboxamido group is important in catalysis (39). The carboxyl group of lactate binds to the quanidinium group of Arg-171.

The chemical mechanism probably involves His-195, acting as a general base. A hydride ion transfer to the 4 position of the nicotinamide ring then occurs. This transfer is stereospecific—that is, the hydride ion is transferred to the *re* face of NAD$^+$ and NADP$^+$. In the reverse reaction, pyruvate reduction, the *pro-R* hydrogen of the nicotinamide ring is transferred:

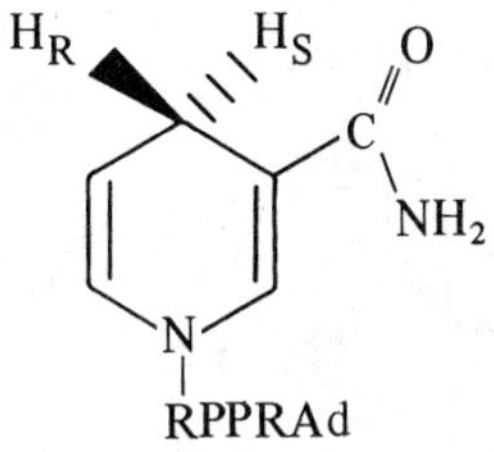

Not all dehydrogenases catalyze hydride transfer to the *re* face: glyceraldehyde-3-P-dehydrogenase is a notable exception. Class A enzymes transfer to the *re* face of NAD$^+$ and use the *pro-R* hydrogen of NADH. Class B enzymes, such as glyceraldehyde-3-P-dehydrogenase, do the opposite. Simon and Kraus have compiled a list of A- and B-type enzymes, part of which appears as Table 4.2 (43).

TABLE 4.2. Examples of "A" and "B" Dehydrogenases[a]

Dehydrogenases	Coenzyme	Type
Glutamate	NAD$^+$ or NADP$^+$	B
Glucose-6-phosphate	NADP$^+$	B
Glyceraldehyde-3-phosphate	NAD$^+$	B
Alcohol	NAD$^+$	A
Lactate	NAD$^+$	A
Isocitrate	NADP$^+$	A

[a]H. Simon and A. Kraus, in *Isotopes in Organic Chemistry*, Vol. 2 (E. Buncel and C. Lee, Eds.) Elsevier, New York (1976), p. 173.

Lysozyme

In 1922 Alexander Fleming, the discoverer of penicillin, accidentally dropped nasal mucus (his own) on a bacterial culture dish. The drop of mucus lysed (dissolved) the bacteria, and the active agent was identified as the enzyme lysozyme.

The enzyme is found in a number of secretions, including tears, saliva, and egg white. Lysozyme can be isolated readily by isoelectric precipitation. Canfield (44) showed that the enzyme was composed of one polypeptide chain of 129 amino acids. Four disulfide bridges were established and the complete sequence was known in 1965. Phillips and collaborators reported a high resolution (2 Å) crystal structure in 1967 (45).

Lysozyme catalyzes the hydrolysis of α-1,4 glycosidic linkages. An excellent substrate is NAG-NAM-NAG-NAM-NAG-NAM:

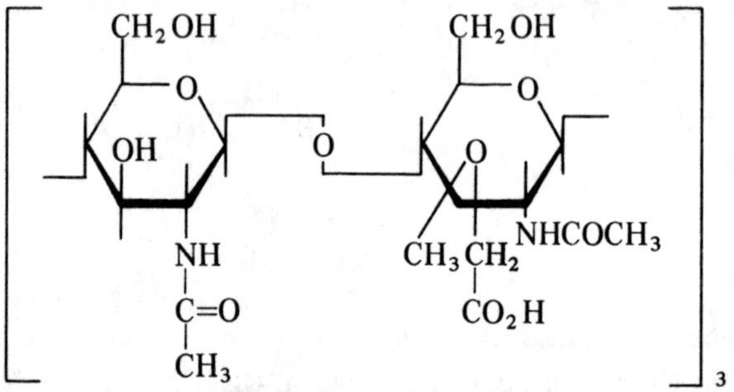

The sugar rings of the hexamer may be labeled A B C D E F; the chain is cleaved between the D and E units. Sub-site binding was proposed on the basis of model building studies; the substrate fits very tightly into a crevice on the enzyme surface.

Crystallographic and solution studies suggest the presence of two carboxyl groups at the active center. The mechanistic scheme shown in Figure 4.11 depicts

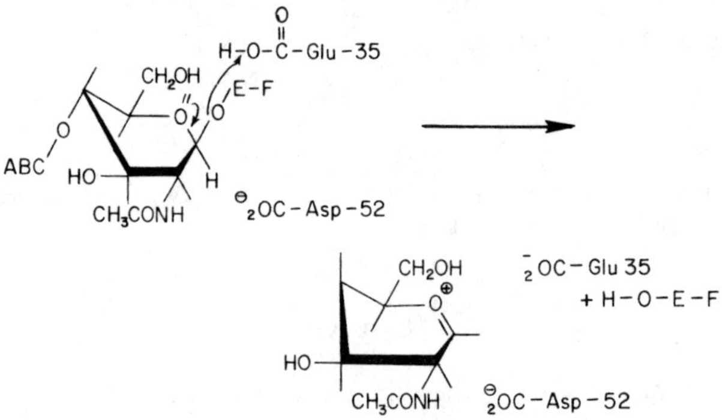

Figure 4.11. The mechanistic roles of glutamic acid-35 and aspartic acid-52 at the active center of lysozyme.

D. EXAMPLES SPECIFIC ENZYME MECHANISMS

TABLE 4.3. Estimated Binding Energies for Sub-Sites: The Complex of Lysozyme and NAG (A) -NAM (B) -NAG (C) -NAM (D) -NAG (E) -NAM (F)

Sub-Site	Residue	Binding Free Energy
A	NAG	−2
B	NAM	−4
C	NAG	−5
D	NAM	+4
E	NAG	−3
F	NAM	−3

the roles played by the carboxyl groups. Glu-35 is in an apolar environment. The γ-carboxylate serves to protonate the leaving glycosidic oxygen. The β-carboxyl group exists in an apolar environment and is therefore shown in the ionized state. The negatively charged carboxylate stabilizes the glycosyloxocarbonium ion intermediate. The existence of the half-chair (or sofa) conformation has been supported by crystallographic (45), NMR (46), and inhibitor-binding studies (47). Mechanical strain (destabilization) has been proposed as a result of unfavorable binding at the D subsite; favorable binding at the other subsites provides the driving force for enzyme-substrate complex formation (Table 4.3).

Lienhard and his collaborators made transition state analogs for lysozyme. One such compound, the lactone of *tetra*-N-acetylchitotetraose, binds to lysozyme more strongly than does *tetra*-N-acetylchitotetraose itself (48). This lactone (below) has a planar sp_2 carbon 1, which resembles the carbon atom of the oxocarbonium ion proposed for lysozyme. Lienhard and coworkers estimated the binding advantage provided by the lactone at the D site to be a factor of 6×10^3. This advantage corresponds to 5 kcal/mol, or a maximum rate enhancement of about 4×10^3.

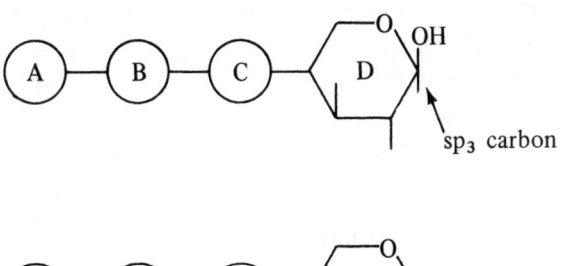

Using refined coordinates of the crystal structure of the enzyme, Levitt has provided conformational energy calculations (49). In contrast to the conclusions mentioned above, he suggests that the substrate binds with all rings in the full-chair conformation. The stabilization of the oxocarbonium ion would still be provided by the carboxylate of Asp-52, however.

E. RATE ENHANCEMENT IN ENZYMATIC REACTIONS

The hydrolysis of urea proceeds with a rate constant of 3×10^{-10} s^{-1} in the absence of a catalyst. In the presence of urease the reaction proceeds much faster—$k = 3 \times 10^4$ sec^{-1}:

$$NH_2 CONH_2 \xrightarrow{k_{uncat}} \text{products} \ (k_{uncat} = 3 \times 10^{-10} s^{-1})$$

$$NH_2 CONH_2 \xrightarrow[\text{urease}]{k_{cat}} \text{products} \ (k_{cat} = 3 \times 10^4 s^{-1})$$

The rate enhancement brought about by the enzyme is 10^{14}. Studies of other enzyme systems provide a range for enzyme-produced rate enhancement of from 10^6 to 10^{14}. The attempts to explain these enormous rate enhancements using quantitative or semi-quantitative arguments have occupied many pages of the literature in the last three decades, and the theories have ranged from simple to exotic. Chemical and crystallographic studies have generally yielded results favoring simple explanations, but this statement does not suggest that enzymes lack sophistication. Enzymes are highly developed molecules from any vantage point. Unusually reactive groups, or "entatic" centers, of enzymes result from very complicated folding of the polypeptide composed of 20 amino acid residues with side chains of every possible polarity. As we saw in the last section, enzymes usually have three or more important groups, fixed in precise spatial relationships, at the active center. Moreoever, the specificity of enzymes (often difficult to separate from rate enhancement discussions) is unique and unmatched by the synthetic catalysts that now exist.

Proximity and Orientation

One simple idea is that enzymes bring reactants close together in a bimolecular reaction, and binding forces bring the substrate into close proximity with a reactive group on the enzyme in a one-substrate reaction. Serine proteases may be

E. RATE ENHANCEMENT IN ENZYMATIC REACTIONS

used as an example. (Water is a second reactant but is also in excess as the solvent.) In 1960, Bruice and Pandit published the data that appear in Table 4.1 (50). These esters hydrolyze via an anhydride intermediate. Much has been said in the past 20 years about the rate enhancements shown in this series of compounds. Bruice and Pandit said in 1960 that the *gem*-dimethyl substituents of 2 and the double bond in compound 4 restrict the number of nonproductive rotational configurations available to the molecules. They said, in essence, that compounds such as 5 hydrolyze very quickly because the reacting groups line up correctly. Since then the following terms and/or theories have appeared (51):

Entropy loss
Propinquity
Rotomer distribution
Orbital steering
Stereopopulation control
FARCE (Freezing at Reactive Centers of Enzymes)
Togetherness

Page and Jencks have argued convincingly that the translational and rotational entropy loss in reactants at the enzyme active center can account for a rate enhancement of 10^8 (52). In brief, a large part of enzymatic rate enhancement results from bringing the reactants into close proximity and aligning the reacting groups properly.

Proximity of Catalytic Groups at the Active Center and Reaction Order

One answer to the question of how enzymes help the substrate along the reaction coordinate to the transition state is that enzymes *are* transition states—minus the substrate. Look at the active center of carboxypeptidase A (Fig. 4.6). At least four components of the active center are believed to be essential in the acylation step. The presence of four critical groups or sites along with the substrate means a five-membered transition state. For a series of displacement reactions Bruice and coworkers have shown that the following empirical formula is applicable in some cases (53):

$$\text{Reaction order} = \frac{-(\text{experimentally determined } T\Delta S)}{4\text{-}5 \text{ kcal/mol}}$$

This means that a change in order of one is accompanied by a $T\Delta S$ requirement of 4-5 kcal/mol. Thus, it is understood that third-order reactions are rare and fifth-order reactions are unknown. To go from a simple second-order displacement to a fifth-order reaction would require a change in $\Delta G^{\ddagger}$ of 3 × 4.5 kcal/mol or 13.5 kcal/mol. The rate enhancement that would result from fixation of four catalytic groups at an enzyme active center would be about 10^6. An analysis of this type is not rigorous. However, such an empirical estimate can establish an upper limit for this kind of rate enhancement contribution. The argument can be extended to other enzymes because multiple catalytic groups exist in all of the enzymes studied to date. But the relative importance of each group at the active center is difficult to judge in quantitative terms.

Destabilization

It is believed that enzymes can use binding energy to bring about a rate enhancement by destabilizing the reactants in relation to the transition state (51). The destabilization may occur in several ways. Geometric destabilization is another way of saying "strain" or "distortion." Destabilization can also be manifested as desolvation or electrostatic effects.

Stretching or twisting of covalent bonds are possible ways by which enzymes use binding energy. Another possible mechanism is the compression of reactants to an internuclear distance that is smaller than the sum of their van der Waals radii. An illustration of this important concept is shown in Figure 4.12.

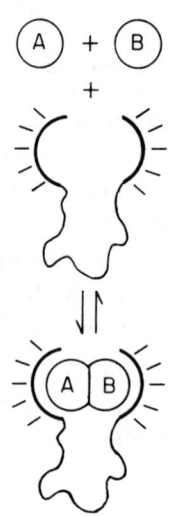

Figure 4.12. Utilization of binding free energy for compression of reactants to an internuclear distance shorter than the sum of the van der Waals radii.

E. RATE ENHANCEMENT IN ENZYMATIC REACTIONS

Effects on the state of solvation of a substrate are probably of great importance. Where substrate binding results in stripping of solvent from an ion in the substrate, very strong activation of the substrate can occur. There are excellent model reactions to support this principle. Crosby and collaborators (54) showed a dramatic solvent dependence for the decarboxylation of a thiamine-pyruvate adduct (1):

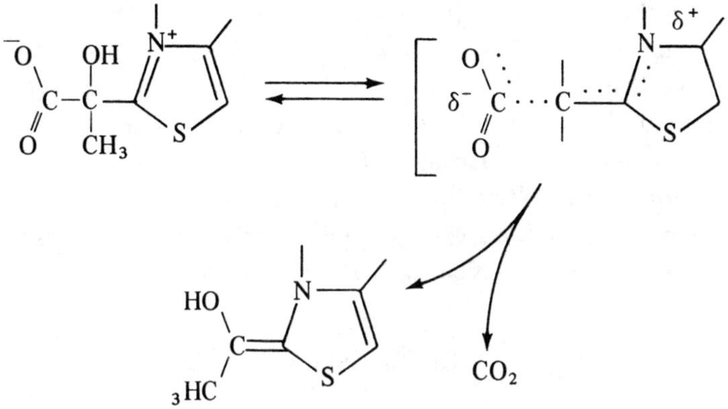

The rate of decarboxylation is 10^4 or 10^5 times faster in ethanol than in water. Transfer from water, a good solvent for carboxylate anion, to a less polar medium would destabilize the reactant relative to the transition state.

Similar medium effects have been demonstrated with nitrobenzisoxazole-carboxylic acid, which decarboxylates 10^8 faster in hexamethylphosphoramide than in water (55):

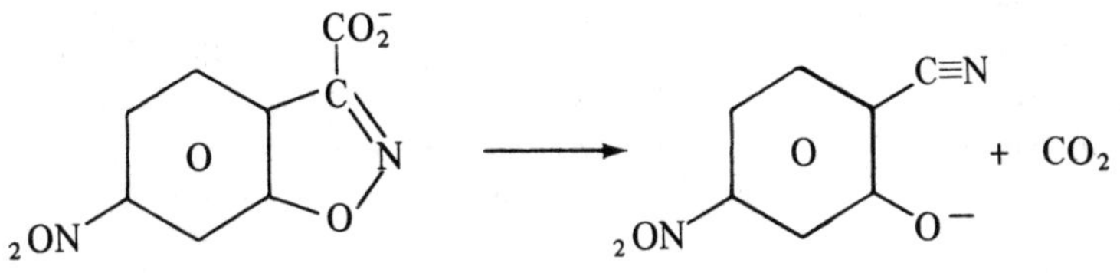

Klotz and coworkers have shown that polyethyleneimine substituted with apolar groups catalyzes the decarboxylation of 2 with a rate enhancement of 1300 over the uncatalyzed reaction in water (56). The system exhibits Michaelis-Menten kinetics with $k_{cat} = 4 \times 10^{-3}$ sec^{-1} and $K_m = 10^{-4}$ M. The mechanism of action of the "synzyme" is to bind 2 into apolar sites, which destablizes it in favor of the charge-delocalized transition state (57).

F. CONCLUSIONS

It is not possible to characterize quantitatively all the contributions to enzymatic rate enhancement discussed here. However, maximum values can be *estimated*. General-acid and general-base catalysis can contribute a factor of 10^2. Covalent catalysis could provide an additional enhancement of 10^3. Translational and rotational entropy loss of reactants and fixation of catalytic groups can give maximum accelerations of about 10^8 and 10^9, respectively. Destabilization can account for an additional factor of 10^3. Obviously, no given enzyme uses all of these possible sources of rate enhancement to the fullest possible extent. It is evident from the listing that the range of enzymatic rate enhancement given earlier (10^6 to 10^{14}) can be accounted for with quite a bit of rate acceleration to spare. The *precise* establishment of the elements of enzyme active centers and their contributions to catalysis could occupy scientists for many years to come. The extent to which this goal is pursued remains to be seen.

REFERENCES

1. T. C. Bruice and S. Benkovic, *Bioorganic Mechanisms,* Benjamin, New York (1966).
2. W. P. Jencks, *Catalysis in Chemistry and Enzymology,* McGraw-Hill, New York (1969).
3. M. L. Bender, *Mechanisms of Homogeneous Catalysis from Protons to Proteins,* Wiley, New York (1971).
4. S. A. Bernhard, *The Structure and Function of Enzymes,* Benjamin, New York (1968), p. 192.
5. M. L. Bender and B. W. Turnquest, *J. Am. Chem. Soc.* 79, 1652 (1957).
6. T. C. Bruice and G. L. Schmir, *J. Am. Chem. Soc.* 79, 1663 (1957).
7. W. P. Jencks and J. Carriuolo, *J. Biol. Chem.* 234, 1272 (1959).
8. B. M. Anderson, E. H. Cordes, and W. P. Jencks, *J. Biol. Chem.* 236, 455 (1961).
9. M. L. Bender, E. J. Pollock, and M. C. Neven, *J. Am. Chem. Soc.* 84, 595 (1962).
10. W. P. Jencks and J. Carriuolo, *J. Am. Chem. Soc.* 83, 1743 (1961).
11. M. L. Bender, G. E. Clement, F. J. Kezdy, and H. Heck, *J. Amer. Chem. Soc.* 86, 3680 (1964).
12. G. A. Hamilton and F. H. Westheimer, *J. Am. Chem. Soc.* 81, 6332 (1959).
13. A. S. Mildvan, *Accts. Chem. Res.* 10, 246 (1977); A. S. Mildvan, *Adv. Enzymol. Rel. Areas Mol. Biol.* 49, 103 (1979).
14. A. S. Mildvan, A. I. Sloan, C. H. Fung, R. K. Gupta, and E. Melamud, *J. Biol. Chem.* 251, 2431 (1976).
15. J. A. Hartsuck and W. N. Lipscomb, in *The Enzymes,* Vol. 3, (P. D. Boyer, Ed.), Academic Press, New York (1971), p. 1.
16. R. M. Stroud, L. M. Kay, and R. E. Dickerson, *J. Mol. Biol.* 83, 185 (1974).
17. H. C. Watson, D. M. Shotton, J. M. Cox, and H. Muirhead, *Nature* 225, 806 (1970).

REFERENCES

18. L. T. Delbaere, W. L. B. Hutcheon, M. N. G. James, and W. E. Thiessen, *Nature* **257**, 758 (1975).
19. M. N. G. James. *J. Philos. Trans. R. Soc. London, Sec. B.* **257**, 119 (1970).
20. J. Drenth, W. G. J. Hol, J. N. Jansonius, and R. Kolkoek, *Eur. J. Biochem.* **26**, 177 (1972).
21. D. M. Blow, J. J. Birktoff, B. S. Harley, *Nature* **221**, 337 (1969).
22. M. W. Hunkapiller, S. H. Smallcombe, D. R. Whitaker, and J. H. Richards, *Biochemistry* **12**, 4732 (1973).
23. J. L. Markley and M. A. Porubcan, *J. Mol. Biol.* **102**, 487 (1976).
24. G. Robillard and R. G. Shulman, *J. Mol. Biol.* **86**, 519 (1974).
25. R. E. Koeppe and R. M. Stroud, *Biochemistry* **15**, 3450 (1976).
26. W. W. Bachovchin and J. D. Roberts, *J. Am. Chem. Soc.* **100**, 8041 (1978).
27. L. W. Harrison, D. S. Auld, and B. L. Vallee, *Proc. Natl. Acad. Sci. USA* **72**, 3930 (1975), and references therein.
28. T. C. Bruice and U. K. Pandit, *J. Am. Chem. Soc.* **82**, 5858 (1960).
29. A. R. Fehrst and A. J. Kirby, *J. Am. Chem. Soc.* **89**, 4857 (1967).
30. T. H. Fife and V. L. Squillacote, *J. Am. Chem. Soc.* **99**, 3762 (1977).
31. M. W. Makinen, L. C. Kuo, J. J. Dymowski, and S. Jaffer, *J. Biol. Chem.* **254**, 356 (1979).
32. M. W. Makinen, K. Yamamura, and F. T. Kaiser, *Proc. Natl. Acad. Sci. USA* **73**, 3882 (1976).
33. R. Breslow and D. L. Wernick, *Proc. Natl. Acad. Sci. USA* **74**, 1303 (1977).
34. D. A. Buckingham, in *Biological Aspects of Inorganic Chemistry,* (D. Dolphin, Ed.), Wiley, New York (1977), p. 141.
35. M. A. Wells and T. C. Bruice, *J. Am. Chem. Soc.* **99**, 5341 (1977).
36. R. Breslow, D. E. McClure, R. S. Broun, and J. Eisenach, *J. Am. Chem. Soc.* **97**, 194 (1975).
37. C. L. Markert and F. Moller, *Proc. Natl. Acad. Sci. USA* **45**, 753 (1959).
38. J. Everse and N. O. Kaplan, *Adv. Enzymol. Rel. Areas Mol. Biol.* **28**, 61 (1973).
39. J. J. Holbrook, A. Lijas, S. J. Steindel, and M. G. Rossman, *The Enzymes,* Vol. 11, 3rd ed., (P. D. Boyer, Ed.), Academic Press, New York (1975), p. 191.
40. D. B. Millar and G. W. Schwert, *J. Biol. Chem.* **238**, 3249 (1963).
41. C. Woenckhaus, J. Berhauser, and G. Pfleiderer, *Hoppe-Seyler's Z. Physiol. Chem.* **350**, 473 (1969).
42. J. Berghauser, I. Falderbaum, and C. Woenckhaus, *Hoppe-Seyler's Z. Physiol. Chem.* **352**, 52, (1971).
43. H. Simon and A. Kraus, in *Isotopes in Organic Chemistry,* Vol. 2 (E. Buncel and C. Lee, Eds.), Elsevier, New York (1976), p. 173.
44. R. Canfield, *J. Biol. Chem.* **240**, 1997 (1965).
45. C. Blake, G. Muir, A. North, D. Phillips, and V. Sarma, *Proc. Roy. Soc. Ser. B.* **167**, 378 (1967).
46. B. D. Sykes, S. L. Patt, and D. Dolphin, *Cold Spring Harbor Symp. Quant. Biol.* **35**, 29 (1971).

47. D. M. Chipman and N. Sharon, *Science* **165**, 454 (1969).
48. I. I. Secemski and G. E. Lienhard, *J. Am. Chem. Soc.* **93**, 3549 (1971); I. I. Secemski, S. S. Leher, and G. E. Lienhard, *J. Biol. Chem.* **247**, 4740 (1972).
49. M. Levitt, in *Peptides, Polypeptides and Proteins,* (E. R. Blout, F. A. Bovey, M. Goodman, and N. Lotan, Eds.), Wiley, New York (1974), p. 99; A. Warshal and M. Levitt, *J. Mol. Biol.* **103**, 227 (1976).
50. T. C. Bruice and U. K. Pandit, *J. Am. Chem. Soc.* **82**, 5858 (1960); T. C. Bruice and U. K. Pandit, *Proc. Natl. Acad. Sci. USA* **46**, 402 (1960).
51. W. P. Jencks, *Adv. Enzymol. Rel. Areas Mol. Biol.* **43**, 219 (1975).
52. M. I. Page and W. P. Jencks, *Proc. Natl. Acad. Sci. USA* **68**, 1678 (1971); M. I. Page, *Chem. Soc. Rev.* **2**, 295 (1973).
53. T. C. Bruice, in *The Enzymes,* Vol. 2, 3rd ed. (P. D. Boyer, Ed.), Academic Press, New York (1970), p. 217.
54. J. Crosby, R. Stone, and G. E. Lienhard, *J. Am. Chem. Soc.* **92**, 2891 (1970).
55. A. S. Kemp and K. Paul, *J. Am. Chem. Soc.* **92**, 2553 (1970); **97**, 7305 (1975).
56. J. Suh, I. S. Scarpa, and I. M. Klotz, *J. Am. Chem. Soc* **98**, 7060 (1976).
57. I. M. Klotz, G. P. Royer, and A. R. Sloniewsky, *Biochemistry* **8**, 4752 (1969).

CHAPTER FIVE

SPECIFICITY

The importance of specificity in biology cannot be overemphasized, and the complexity of enzyme structure is necessary to provide it. Enzyme specificity for a given reaction can be defined as the selection of one substrate over another.

The multipoint mode of substrate binding provides not only strength, but the specific orientation required for biological catalysis. A specific or good substrate reacts at a rate that is *relatively* rapid compared to that of a poor substrate of the same enzyme.

There are many vantage points from which to view specificity. We shall consider reaction specificity, structural specificity, binding specificity, and catalytic specificity. Many aspects of these topics are connected.

A. REACTION SPECIFICITY

Enzymes have been classified into six major reaction groups by the International Union of Biochemistry (1).

1. *Oxidoreductases.* These enzymes catalyze redox reactions, resulting in oxygenation or overall removal or addition of hydrogen atom equivalents.

2. *Transferases.* Enzymes of this group catalyze the transfer of one-carbon groups, acyl, sugar, phosphoryl, and sulfur-containing groups.

3. *Hydrolases.* The broad range of substrates hydrolyzed by enzymes in this group includes esters, anhydrides, peptides, and glycosides.

4. *Lyases*. Enzymes in this class catalyze formation of or additions to double bonds such as C=C, C=O, and C=N.

5. *Isomerases*. A considerable number of isomerizations, such as epimerization, can be catalyzed by members of this group.

6. *Ligases*. These enzymes catalyze bond formation (with concomitant ATP cleavage).

There is some flexibility in the definition of reaction specificity. A hydrolase may function as a catalyst for peptide hydrolysis, esterolysis, and aminolysis reactions. In this regard chymotrypsin could be termed an acyl transferase, rather than a hydrolase. The *natural* catalytic function of the enzyme is, however, hydrolysis of peptide bonds.

Limitation of reaction specificity is a basic requirement for a biological system. This restriction permits the operation of many enzymes within a cell as catalysts of sequential and concurrent reactions in a controlled manner, without interference of one enzyme with another.

B. STRUCTURAL SPECIFICITY

Examples of structural specificity are shown in Figure 5.1. Trypsin catalyzes the hydrolysis of N-α-benzoylarginine ethyl ester 1000 times faster than the hydrolysis of acetylglycine ethyl ester. The hydrolysis is regiospecific (2). In a natural protein substrate, trypsin cleaves on the carboxyl side of arginine and lysine. Regiospecificity in enzyme-catalyzed reactions is an important consideration in the use of enzymes in organic synthesis. This advantage has been exploited by Isowa and collaborators (3) in the enzymatic synthesis of aspartame (aspartylphenylalanine methyl ester), which is a nutritive sweetener. Thermolysin, normally a protease, can catalyze amide bond formation at pH 6 to 8 in the presence of excess amine. Cbz-Asp can be condensed in a regiospecific manner with Phe-OMe:

$$\text{Cbz-NHCH(CH}_2\text{CO}_2\text{H)CO}_2\text{H} + 2\,\text{HNCH(CH}_2\text{Ph)CO}_2\text{Me} \xrightleftharpoons{\text{thermolysin}} \text{Cbz-NHCH(CH}_2\text{CO}_2\text{H)CONHCH(CH}_2\text{Ph)CO}_2\text{Me}$$

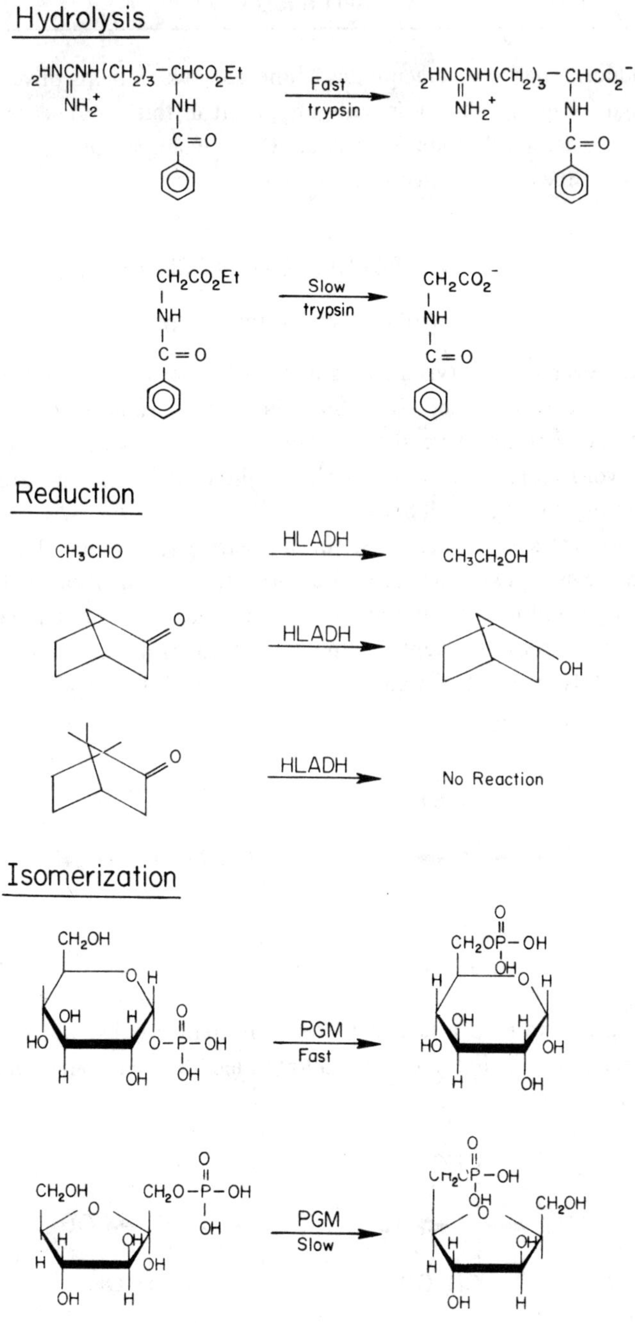

Hydrolysis

$_2HNCNH(CH_2)_3-CHCO_2Et$ →(Fast trypsin)→ $_2HNCNH(CH_2)_3-CHCO_2^-$
 ‖ |
NH_2^+ NH
            C=O
            Ph

CH_2CO_2Et—NH—C(=O)—Ph →(Slow trypsin)→ $CH_2CO_2^-$—NH—C(=O)—Ph

Reduction

CH_3CHO →(HLADH)→ CH_3CH_2OH

norcamphor →(HLADH)→ norbornanol

camphor →(HLADH)→ No Reaction

Isomerization

glucose-1-phosphate →(PGM Fast)→ glucose-6-phosphate

ribose-1-phosphate →(PGM Slow)→ ribose-5-phosphate

Figure 5.1. Structural specificity.

The product, aspartame, precipitates as the salt, which helps drive the reaction to the right. Regiospecificity is quite important in this case because the protection of the β-carboxyl group is obviated. Other examples of the use of enzymes in organic synthesis are treated in Chapter 7.

C. STEREOCHEMICAL SPECIFICITY

Notation and Terminology

Pasteur discovered two crystal forms of tartaric acid: one caused the rotation of plane-polarized light to the right, the other caused rotation to the left. Also, Pasteur found a strain of yeast that would ferment the dextrorotatory form but not the levorotatory form. In other words, there are systems in nature that possess the ability to distinguish between optical isomers.

For many years after Pasteur, stereochemistry was dominated by the object-mirror image test: two isomers of a compound were enantiomers if they were mirror images and not superimposable. Compounds with an asymmetric carbon (one to which four different groups are attached) are common examples in biology. D,L-isomers of alanine are shown here with the aid of perspective formulae.

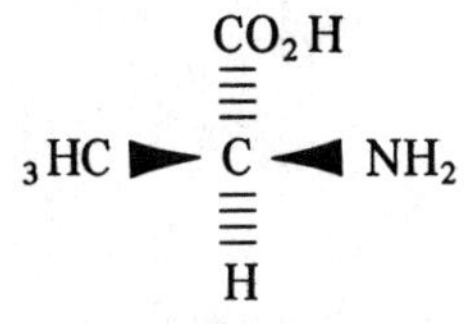

$$_3HC \blacktriangleright \underset{H}{\overset{CO_2H}{C}} \blacktriangleleft NH_2 \qquad _2HN \blacktriangleright \underset{H}{\overset{CO_2H}{C}} \blacktriangleleft CH_3$$

L-alanine 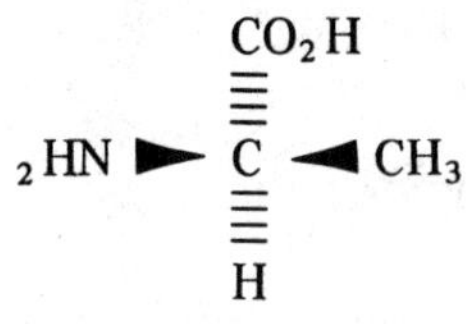　　　　　　　D-alanine

The α-carbon atom is attached to four different groups, which makes it an asymmetrical (or "chiral") center. Glyceraldehyde enantiomers may be illustrated similarly.

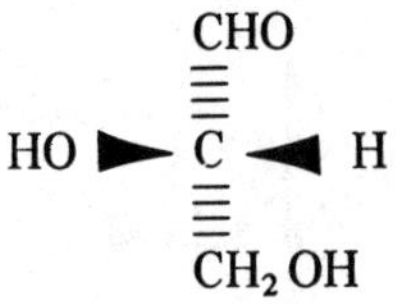

$$HO \blacktriangleright \underset{CH_2OH}{\overset{CHO}{C}} \blacktriangleleft H \qquad H \blacktriangleright \underset{CH_2OH}{\overset{CHO}{C}} \blacktriangleleft OH$$

L-glyceraldehyde 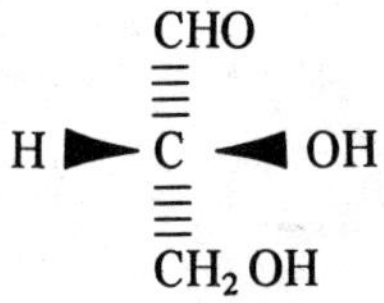　　　　D-glyceraldehyde

C. STEREOCHEMICAL SPECIFICITY

D,L notation and the terms erythro, threo, myo, allo, and so on are useful within a series of compounds that are related in some way (4, 5). The amino acids and aldohexose sugars, for example, may be described in D,L notation. The serious drawback associated with these terms is that they are relative, rather than absolute indicators of configuration. The R,S descriptions of Cahn, Ingold, and Prelog have supplanted traditional notation in most areas (6). The R,S system is more convenient in that it denotes absolute, rather than relative asymmetry. The groups attached to the asymmetric center are assigned a priority related to their mass, and so on (6). Groups found frequently in nature follow in decreasing order of priority: -SH > -OR > -OH > -NHCOR > -NH$_2$ > -CO$_2$R > -CHO > -CH$_2$OH > -C$_6$H$_5$ > -CH$_3$ > -T > -D > -H. Assignment of the R or S designation to the chiral center is done in two steps. First, the groups or atoms are given priority ranking. For alanine the priority order is -NH$_2$ > COOH > -CH$_3$ > H. Next, the molecule is viewed along the axis formed by the chiral center and the group of lowest priority from the direction of the chiral center. L-alanine looks like this:

$_2$HN — C — CH$_3$ Direction of decreasing order of priority

CO$_2$H

S-alanine

The hydrogen atom (the substituent of the lowest priority) is behind the chiral carbon. The order of priority of the ligands other than —H decreases in a counterclockwise direction, which means an S designation applies. D-alanine looks like this when viewed from a direction opposite the ligand of lowest priority:

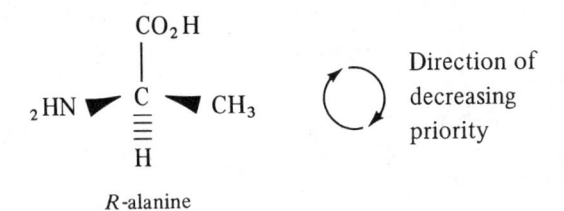

CO$_2$H
|
$_2$HN — C — CH$_3$ Direction of decreasing priority
H

R-alanine

The direction of decreasing priority in this case is clockwise, which means an R designation applies. Starting at 12 o'clock, if the direction of decreasing order is to the right (rectus), the R designation applies. If the direction of priority order decrease is to the left (sinister), the S designation applies.

Another important aspect of stereochemical distinction in biology involves the discrimination between like groups attached to a "prochiral" center. Consider the central carbon of citric acid.

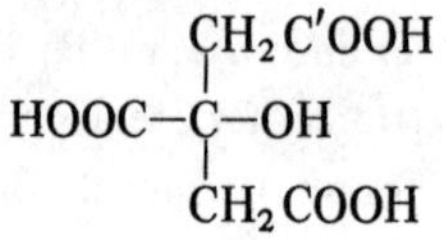

$$\begin{array}{c} CH_2C'OOH \\ HOOC-C-OH \\ CH_2COOH \end{array}$$

This carbon atom is symmetrical because the $-CH_2COOH$ ligands are identical (the prime is for notation only). These groups are also termed homomorphic since, in isolation, they are superimposable (7). It was Ogston, in his classic paper of 1948, who showed that the $-CH_2COOH$ groups were distinguishable when citrate was bound to an enzyme with three points of interaction (8). View citric acid as shown at the top of Figure 5.2. The binding mode in 5.2a is most favor-

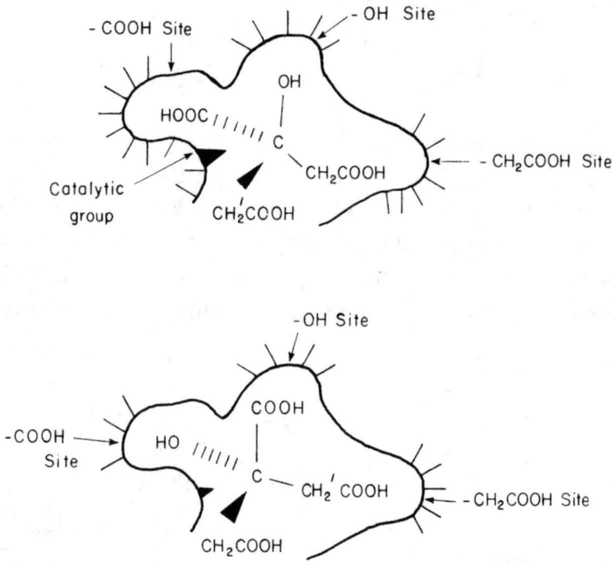

Figure 5.2. Stereoheterotopic specificity. The reactive complex is shown in (a). The non-productive complex (b) may form, but the reactant is in the wrong orientation with regard to the catalytic group. As a result, the asymmetry of the enzyme is imposed on the symmetrical reactant.

C. STEREOCHEMICAL SPECIFICITY

able and the $-CH_2COOH$ group is placed near the catalytic group(s) of the enzyme. Note that, to place the $-CH_2COOH$ group near the catalytic area of the enzyme surface, two of three favorable binding interactions are lost.

Ogston thus showed that an asymmetrical enzyme surface could confer asymmetry on a symmetrical substrate. Bentley points out that Ogston's contribution was critical in the development of prochirality theory (9). Hirschman (7) and Hansen (7) addressed the rotational asymmetry of a molecule, Caa'bc (such as citrate). If one of the groups is replaced by a group, d, the product, Cabcd, is chiral. Hansen termed C a "prochiral" center. Groups a and a' may be termed *pro-R* and *pro-S* as follows [see (5) and original papers cited therein]. For the citric acid molecule,

$$\text{HOOCCH}_2-\overset{\overset{\displaystyle CH_2C'OOH}{|}}{\underset{\underset{\displaystyle COOH}{\blacktriangledown}}{C}}||||\ OH$$

one of the homomorphic groups, $-CH_2COOH$, is arbitrarily given a higher priority ranking ($CH_2'COOH > CH_2COOH$). In general terms

$$a-\overset{\overset{\displaystyle a'}{|}}{\underset{\underset{\displaystyle b}{\blacktriangledown}}{C}}||||\ c$$

if the priority sequence of c > b > a' describes a clockwise arc when viewed from the tetrahedral center opposite the ligand of lowest priority, a' is designated *pro-R* or a_R. Alternatively, if c → b → a' priorities decrease in the counterclockwise direction, a' is specified *pro-S*. Let us illustrate the system with citrate (Fig. 5.3). First, assign priority ratings using the usual perspective formula (Fig. 5.3). The $-CH_2COOH$ group at the top is arbitrarily given priority over the group at the bottom. The order is then $-OH > -COOH > -CH_2C'OOH > -CH_2COOH$. The molecule is rotated so that the ligand of lowest priority, $-CH_2COOH$, is placed back into the paper (Fig. 5.3). The priority rankings (in parentheses) of the projecting groups decrease in a clockwise fashion, which means the $-CH_2C'OOH$ is *pro-R*. If the bottom $-CH_2-COOH$ group of the perspective formula (left, Fig. 5.3) is given top priority and the same analysis is carried out, it will be designated *pro-S*. Obviously, once one of the *pro-R* or *pro-S* assignments is made, the other follows directly. A good check, however, is to make both assignments independently.

SPECIFICITY

Priority order: $-OH > -COOH > -CH_2C'COOH > -CH_2COOH$

$$\text{(3) HOOC} - \overset{CH_2\overset{'}{C}OOH\ (2)}{\underset{CH_2COOH\ (1)}{C}} - OH\ (4)$$

$$\text{(3) HOOC} \underset{HO\ (4)}{\overset{CH_2\overset{'}{C}OOH\ (2)}{C}} CH_2COOH\ (1)$$

(3) ⟶ (2) ⟶ (4) ∴ $CH_2'COOH$ is <u>Pro</u>-R

Figure 5.3. Assignment of the *pro-R* designation in citrate.

The carboxymethyl groups of citrate are termed *stereoheterotopic*. Although they are structurally identical when isolated, they are distinguishable ligands of citrate in the presence of an assymetrical agent such as an enzyme. The faces of a compound containing a trigonal carbon atom are also stereoheterotopic. The faces are termed *re* and *si*, as follows. Consider the molecule C_{abc}. It can be viewed from above or below the plane formed by the group of atoms:

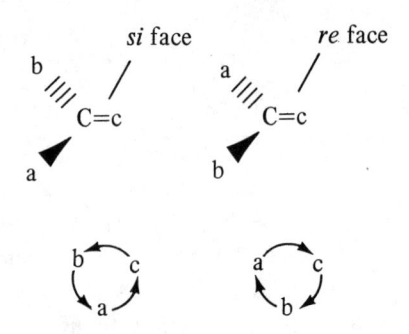

$$\overset{si\ face}{\underset{a}{\overset{b}{C}}=c} \qquad \overset{re\ face}{\underset{b}{\overset{a}{C}}=c}$$

With a priority order of $c > b > a$, the face is *si* if $c \to b \to a$ decreases in a counterclockwise manner. Conversely, the face is *re* if the direction of decreasing priority is clockwise. Acetaldehyde may serve as an example (view from the top).

$$\overset{si}{\underset{H}{\overset{_3HC}{C}}=O} \qquad \overset{re}{\underset{_3HC}{\overset{H}{C}}=O}$$

C. STEREOCHEMICAL SPECIFICITY

For olefins there are two trigonal centers, and the face will be designated *re-re*, *re-si*, and so on:

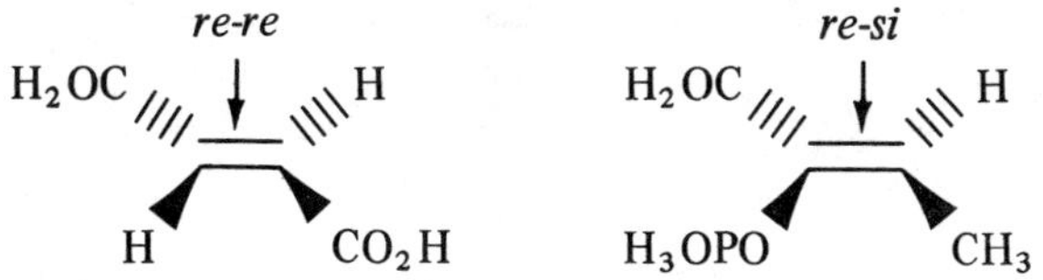

re-re

H₂OC ⫽⫽⫽ ↓ ⟍⟍ H
H   CO₂H

re-si

H₂OC ⫽⫽⫽ ↓ ⟍⟍ H
H₃OPO   CH₃

Examples of Stereospecificity in Enzyme-Catalyzed Reactions

The binding and catalytic sites of α-chymotrypsin are shown in Figure 5.4 with acetyl-L-phenylalanine methyl ester and acetyl-D-phenylalanine methyl ester. The L-enantiomer is the specific substrate. The D form binds, and in fact the acyl enzyme can form, but enzyme deacylation is extremely slow. The distinction between *R* and *S* enantiomers may be illustrated in many systems. This stereospecificity is quite important from a practical point of view. As was mentioned earlier in this chapter (see also Chapter 7), enzymes can be used in asymmetric synthesis. Moreover, resolution of racemic mixtures, which are products of chemical synthesis, is of industrial importance.

The knowledge of the stereochemistry of substrates and products can provide insight into the mechanism and structure of intermediates of enzyme-catalyzed reactions. Rose and O'Connell used this approach for aldose-ketose isomerization systems, for which they proposed an enediol intermediate (10). Consider a tritiated aldose that is *R* at C-2. If the ketose product is *R* at C-1, the enediol intermediate must be *cis*:

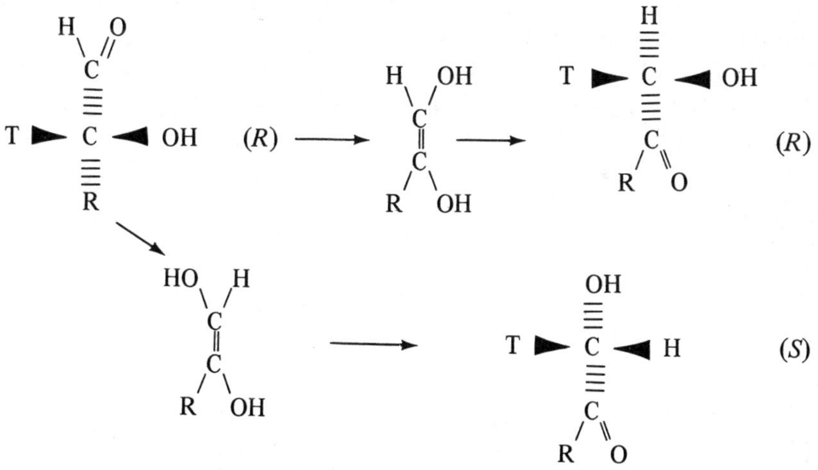

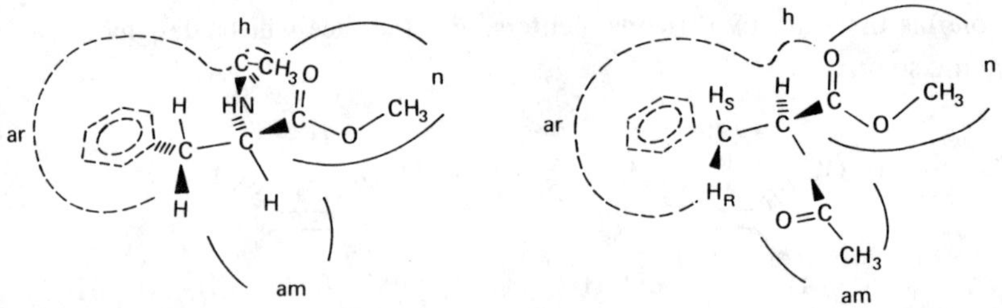

Figure 5.4. Enantiomeric specificity, as shown with acetyl-phenylalanine methyl ester and α-chymotrypsin. The *ar* site is for the aromatic ring; the *am* site is for the *N*-acyl group. The *n* site cannot accommodate the acetyl group [D-enantiomer (*a*)]. The L-enantiomer will bind as shown in (*b*). From J. B. Jones, *Applications of Biochemical Systems in Organic Chemistry*, Techniques in Chemistry Series, Vol. 10, Wiley, New York (1976), p. 137.

The lower route yielding a ketose that is *S* at C-1 would require a *trans* enediol intermediate.

Frey and his collaborators have studied the stereochemical course of catalysis by phosphotransferase enzymes (11, 12). Their work involves the use of ATPγS, $\gamma^{18}O$ with a chiral $\gamma-[^{18}O]$ phosphorothioate group of known configuration:

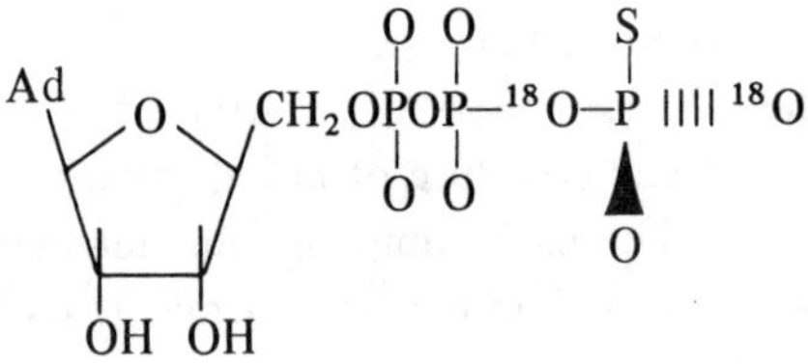

The transfer of the phosphorothioate group can occur with retention or inversion of configuration. Observation of retention is consistent with a single intermediate or odd number of them. Inversion indicates a direct displacement. In other words, retention of configuration is consistent with a double-displacement pathway, such as ping-pong, and inversion is consistent with a single displacement mechanism, such as an ordered mechanism.

Bryant and Benkovic have studied the stereochemical course of snake venom phosphodiesterase, using ATPαS (13):

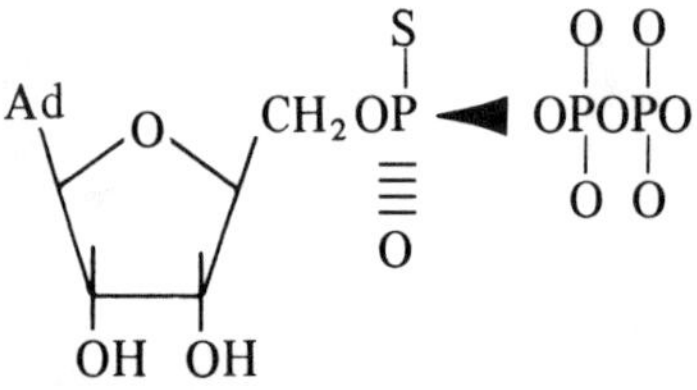

D. BINDING AND CATALYTIC SPECIFICITIES, LOCK AND KEY, INDUCED FIT, AND WRONG-WAY BINDING

Analysis of the configuration of the product AMPαS showed that a single covalent phosphoryl-enzyme intermediate is probably involved.

Emil Fischer proposed the lock-and-key hypothesis to explain specificity in enzyme-catalyzed reactions. The basic idea is that the active center of the enzyme conforms exactly to the shape of the substrate. The enzyme cannot bind substrates larger than the specific substrate. Small substrates could bind, but not so strongly as the specific substrate. Indeed, the observations of crystallographers studying enzyme-pseudosubstrate complexes have indicated a very precise and tight fit of the substrate into the binding site. The difficulty with the lock-and-key theory is that it does not account for catalytic specificity in the sense that good catalysis and good binding are not always related. This problem is illustrated in Table 5.1 (14). Pepsin catalyzes the hydrolysis of peptide esters with a wide range of k_{cat} values (0.0025-2.43 sec^{-1}). The K_m value, which is essentially a K_D, varies by a factor of only 4. The substrate CBZ-His-Phe-Leu-OMe binds better than the substrate CBZ-Gly-His-Phe-OEt, but the rate of hydrolysis is about 1000 times slower. Good binding is not always accompanied by fast catalytic turnover.

The lock-and-key theory falls short in other applications too. Phosphoglucomutase catalyzes the interconversion of glucose-1-phosphate and glucose-6-phos-

TABLE 5.1. Hydrolysis of Peptides esters by Pepsin at pH 4.0[a]

Peptide[b]	K_m, mM	k_{cat}, sec^{-1}
Cbz-Gly-His-Phe-Phe-OEt	0.80	2.43
Cbz-His-Phe-Trp-OEt	0.23	0.51
Cbz-His-Phe-Trp-OEt	0.18	0.31
Cbz-His-Phe-Tyr-OEt	0.23	0.16
Cbz-His-Tyr-Phe-OMe	0.68	0.013
Cbz-His-Tyr-Tyr-OEt	0.24	0.0094
Cbz-His-Phe-Leu-OMe	0.56	0.0025

[a] Data from K. Inouye and J. S. Fruton, *Biochemistry* 6, 1765 (1967).
[b] Cbz = carbobenzoxy; Gly = glycine; His = histidine; Phe = phenylalanine; Trp = tryptophan; Leu = leucine; Tyr = tyrosine.

phate. The reaction involves glucose-1,6-diphosphate and a phosphorylenzyme intermediate:

$$E - P + Glu\text{-}1\text{-}P \rightleftharpoons E + Glu\text{-}1,6\ DiP$$

$$E + Glu\text{-}1,6\ DiP \rightleftharpoons E - P + Glu\text{-}6\text{-}P$$

The phosphorylenzyme may be prepared by incubation of the free enzyme with glucose-1,6 diphosphate. Ray and coworkers have studied the rate of phosphoryl group transfer from the phosphoryl enzyme to water and alcohol receptors (15). Phosphate transfer to water is 3×10^{10} times slower than transfer to the 6-hydroxyl group of glucose-1-phosphate. Such a large difference in rate cannot be explained by a difference in nucleophilic reactivity of the hydroxyl group of water and the 6-hydroxyl group of glucose-1-phosphate. It is also of interest that the binding of xylose-1-phosphate stimulates the rate of phosphate transfer to water by a factor of 1.7×10^5.

A related example involves the hexokinase reaction. Hexokinase catalyzes the transfer of phosphate from ATP to glucose. Transfer of the phosphate group to water is also catalyzed by hexokinase, but at a rate 5×10^{-6} times as fast as the transfer to the specific substrate, glucose (16).

The enzyme trypsin effectively catalyzes the hydrolysis of benzoyl-arginine ethyl ester (**1**). Acetylglycine ethyl ester (**2**) is hydrolyzed slowly.

$$\underset{\mathbf{1}}{{}_2HN\overset{\overset{\overset{+}{N}H_2}{\|}}{C}NH(CH_2)_3\underset{\underset{\underset{R}{C=O}}{NH}}{C}HCO_2Et} \qquad \underset{\mathbf{2}}{\underset{\underset{\underset{R'}{C=O}}{NH}}{C}H_2CO_2Et}$$

Inagami and Murachi found that the rate of acetylglycine ethyl ester hydrolysis was enhanced by a factor of 7 in the presence of methyl guanidine, a part of the specific substrate (**1**) (17). Inagami and Hatano observed a six-fold increase in the rate of carboxamidomethylation of the active-site histidine when methyl guanidine was present in the reaction solutions (18).

The foregoing examples illustrate specificity manifested in the catalytic step of enzyme reactions. There are two popular theories that explain these effects—the induced-fit theory and the wrong-way binding theory. In the induced-fit hypothesis of Koshland, the free enzyme exists in an inactive state (19). When a

D. BINDING AND CATALYTIC SPECIFICITIES

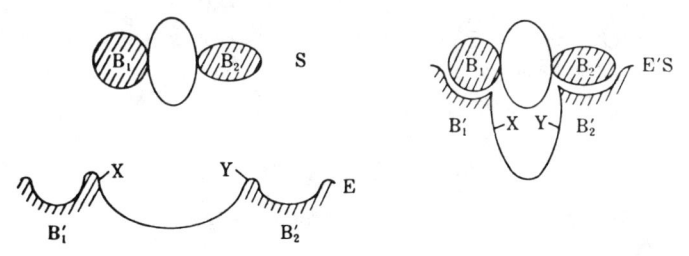

Figure 5.5. The induced-fit model for specificity. Binding free energy is expended to reposition groups X and Y to a reactive state in ES'.

specific substrate binds, the enzyme undergoes a conformational change driven by the substrate binding free energy. The catalytic groups are repositioned in such a way that the enzyme becomes activated (Fig. 5.5). A poor substrate may bind strongly, but not in a way that brings about the conformational change that results in repositioning of catalytic groups. Binding free energy is used in this case for specificity, rather than rate enhancement. The thermodynamic considerations of this model have been treated skillfully by Jencks (20).

The induced-fit model would explain the catalytic specificity shown in the examples of pepsin, phosphoglucomutase, hexokinase, and trypsin. Phosphate transfer from phosphoryl phosphoglucomutase is more rapid with glucose-1-phosphate as acceptor than with water as acceptor. Although water can penetrate to the active site of the enzyme, the catalytic groups are not properly positioned for the phosphate transfer to occur. The induced-fit explanation suggests that glucose-1-phosphate and other ligands induce a conformational change, which results in fast transfer of the phosphate group to water or other hydroxylic acceptors.

The wrong-way binding theory depicts a rigid enzyme. A good substrate possessing several specificity determinants binds strongly to the enzyme in only one possible orientation (Fig. 5.6). The ES complex is always a productive one in that products can be formed. A poor substrate would not have the same number of specificity determinants as the specific substrate. The poor substrate may bind well, but some of the complexes will be nonproductive (Fig. 5.6). Proteolytic enzymes that catalyze the hydrolysis of substrates of the type

$$R^3 \overset{O}{\underset{H}{\overset{\|}{C}}} N \blacktriangleright \overset{H}{\underset{CO_2 R^1}{\overset{\|}{\underset{\|}{C}}}} \blacktriangleleft R^2$$

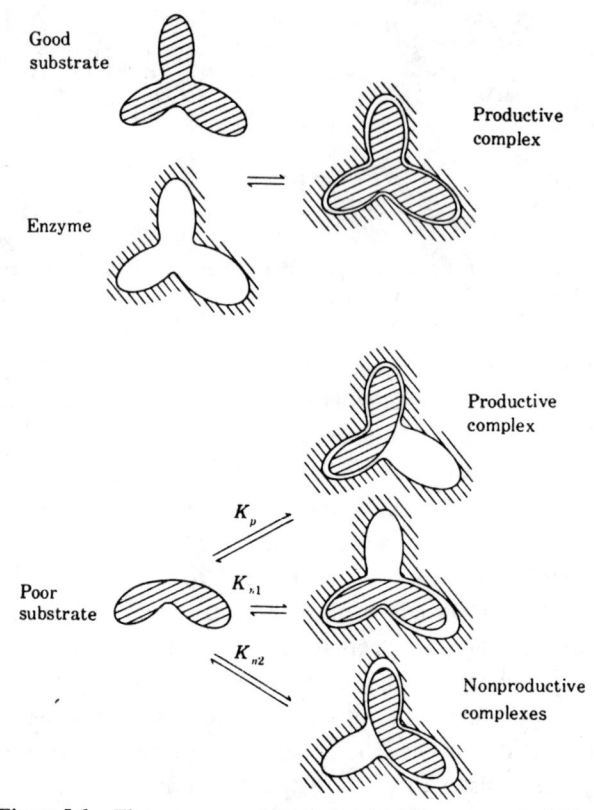

Figure 5.6. The wrong-way binding model. The enzyme is rigid.

can often bind both R and S enantiomers, but only the S enantiomer undergoes hydrolysis. The R form may bind strongly, but in an inverted fashion.

It is not easy to distinguish experimentally between the induced-fit theory and the wrong-way binding hypothesis. Arguments can be made in such specific cases as the hexokinase system. Jencks points out that if the wrong-way binding theory were applicable, water would bind to hexokinase incorrectly 2×10^5 times more often than it binds correctly (21). This binding or orientation requirement seems far too demanding in view of those for other chemical reactions.

A basic difference between the two specificity theories is that the induced-fit model requires a conformational change of catalytic importance and the wrong-way binding theory connotes a rigid enzyme. In the case of hexokinase there is evidence for a substrate-induced conformational change (22 and references therein). The McDonald group showed that glucose induced a conformational change in yeast hexokinase. Using small-angle X-ray scattering, these investigators showed that the radius of gyration of the glucose-enzyme complex was smaller than that of the free enzyme. It has been suggested that the glucose-binding site resides in a deep cleft in the hexokinase molecule, and that the cleft closes when the substrate binds. It is implied that this substrate-induced conformational change results in the repositioning of catalytic groups in such a way that the enzyme is activated.

E. LIMITATIONS OF ENZYME SPECIFICITY

Within a given reaction class, substrate specificity can be broad. The active centers of some dehydrogenases can accommodate a variety of substrate structures. Horse liver alcohol dehydrogenase, for example, will catalyze effectively the reduction of **3**, which is quite different from its natural substrate, acetaldehyde (23).

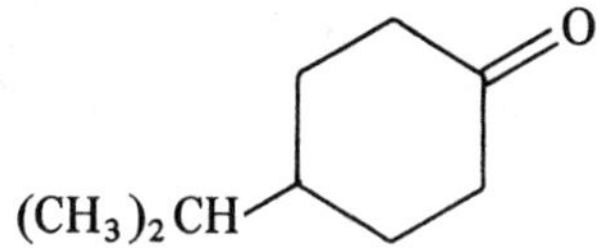

Carboxypeptidase Y, an exopeptidase from yeast, has a very broad specificity. All natural amino acids can be released at detectable rates (24). Synthetic derivatives such as trifluoroacetyllysine and S-*t*-butyl cysteine are also released (25). Other examples include bacterial esterases.

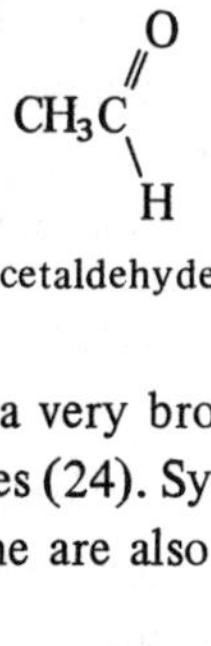

Demands for extremely limited specificity of enzyme reactions are common in biology. Consider for instance, protein biosynthesis. The error rate for misincorporation of valine for isoleucine in about 1 in 3000. This fidelity is quite remarkable in view of the similar structures of valine and isoleucine:

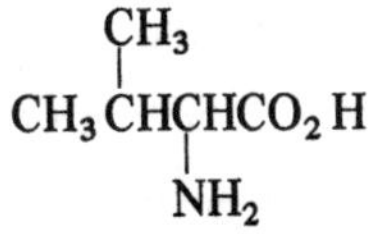

$$CH_3CHCHCO_2H \atop NH_2$$
Valine

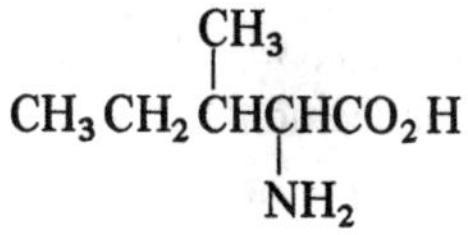

$$CH_3CH_2CHCHCO_2H \atop NH_2$$
Isoleucine

Fersht and Dingwall have provided evidence for the "double-sieve" editing model to explain such accurate fulfillment of exacting specificity demands (26). The editing occurs at the level of aminoacyl t-RNA synthetases. These enzymes catalyze the esterfication of an amino acid with the appropriate transfer RNA molecule. The resultant aminoacyl t-RNAs are used directly for protein biosynthesis. "Mischarging" of the t-RNA would therefore lead to an erroneous incorporation of an amino acid into the linear sequence. The "double-sieve" editing model is based on the observation that a given synthetase has a separate hydrolytic site. Should the t-RNA be misacylated, the ester bond would be hydrolyzed before the aminoacyl-t-RNA would be released into solution:

$$^{Val}E \cdot Ileu \cdot ATP \rightarrow\ ^{Val}E \cdot Ileu\text{-Amp} + PPi$$

$$^{Val}E \cdot Ileu\text{-AMP} + t\text{-RNA}^{Val} \rightarrow\ ^{Val}E \cdot Ileu\text{-}t\text{-RNA}^{Val} + AMP$$

$$^{Val}E \cdot Ileu - t\text{-RNA}^{Val} \rightarrow\ ^{Val}E + Ileu + t\text{-RNA}^{Val}$$

In the first two steps the valyl-t-RNA synthetase charges isoleucine erroneously into t-RNA^{Val}. However, this error is corrected by proofreading at the hydrolytic site. The two sites act as sieves in that the synthetic site rejects amino acids larger than the one for which it is specific; then the hydrolytic site destroys products made from incorrect amino acids that are smaller than the specific amino acid or of similar size but incorrect structure.

F. CONCLUSION

There are examples of specificity in catalysis with nonbiological materials. The discrimination among substrates in these examples is often based on exclusion of molecules larger than the desired reactant. These catalysts might be considered "single-sieve" catalysts. The unique aspect of specificity in enzymes is that molecules smaller than the substrate, as well as molecules of similar size, can be excluded. This trait is evident in the "double-sieve" editing model and in the induced-fit theory. In both cases, specificity does not come without cost. ATP is consumed when mischarged t-RNAs are hydrolyzed. In the induced-fit model, binding free energy, which might normally go into destabilization of the substrate (rate enhancement) is used to induce a catalytically competent conformation of the enzyme. In the complex milieu of the cell, such an investment of energy is a basic necessity to maintain order.

REFERENCES

1. T. Barman, *Enzyme Handbook*, Vol. 1, Springer-Verlag, New York (1969).
2. A. Hassner, *J. Org. Chem.* **33**, 2684 (1968).
3. Y. Isowa, T. Ichikawa, and M. Ohmori, *Bull. Chem. Soc., JPN,* **50**, 2766 (1978).
4. R. Bentley, *Molecular Asymmetry in Biology,* Vols. 1 and 2, Academic Press, New York (1969 and 1970).
5. J. B. Jones, in *Applications of Biochemical Systems in Organic Synthesis, Techniques of Chemistry* series, Vol. 10, Wiley, New York (1976), p. 479.
6. R. S. Cahn, C. K. Ingold, and V. Prelog, *Angew. Chem. Int. Eng. Ed.* **5**, 385 (1966).
7. H. Hirschmann and K. R. Hanson, *Tetratedron* **30**, 3649 (1974).
8. A. G. Ogston, *Nature* **162**, 963 (1948).
9. R. Bentley, *Nature* **276**, 673 (1978).
10. I. A. Rose and E. L. O'Connell, *Biochem. Biophys. Acta* **42**, 159 (1960).
11. J. P. Richard, H. T. Ho, and P. A. Frey, *J. Am. Chem. Soc.* **100**, 7756 (1979).
12. J. P. Richard and P. A. Frey, *J. Am. Chem. Soc.* **100**, 7757 (1979).
13. F. R. Bryant and S. J. Benkovic, *Biochemistry* **18**, 2825 (1979).
14. K. Inouye and J. S. Fruton, *Biochemistry* **6**, 1765 (1967).
15. W. J. Ray and J. D. Owens, *Biochemistry* **15**, 4006 (1976).
16. K. A. Trayser and S. P. Colowick, *Arch. Biochem. Biophys.* **94**, 161 (1961).
17. T. Inagami and T. Murachi, *J. Biol. Chem.* **239**, 1395 (1964).
18. T. Inagami and H. Hatano, *J. Biol. Chem.* **244**, 1176 (1969).
19. D. E. Koshland, Jr. and K. E. Neat, *Ann. Rev. Biochem* **37**, 359 (1967).
20. W. P. Jencks, *Adv. Enzymol. Rel. Areas Mol. Biol.* **43**, 219 (1975).
21. W. P. Jencks, *Catalysis in Chemistry and Enzymology*, McGraw-Hill, New York (1969), p. 291.
22. R. C. McDonald, T. A. Steitz, and D. M. Engleman, *Biochemistry* **18**, 338 (1979).
23. J. M. H. Graves, A. Clark, and H. J. Ringold, *Biochemistry* **4**, 2655 (1965).
24. R. Hayoski, *Methods Enzymol.* **45**, 568 (1976).
25. H. Hsiao, G. M. Ananthramaiah, and G. P. Royer, unpublished data.
26. A. R. Fersht and C. Dingwall, *Biochemistry* **18**, 2627 (1979).

CHAPTER SIX

CONTROL OF ENZYME ACTION

The concentration level of a given enzyme within a cell is determined by its rates of synthesis and degradation. Protein biosynthesis and its control has been an exciting and fruitful research area for the past 25 years (1). Protein degradation is not so well understood (2). A number of important questions remain. What types of proteases are responsible for enzyme degradation? How are they turned on and off? What roles do lysosomal proteases play in protein catabolism? These and other questions are being addressed in a number of laboratories around the world. Obviously, the amount of enzyme activity in a cell depends upon the level of enzyme and the catalytic competence of the enzyme. In this chapter we address the question, *How are existing enzyme molecules turned on and off in response to changes in the environment and metabolic state of the organism?*

Mechanisms for control of enzyme action can be categorized roughly as noncovalent-reversible, covalent-reversible, and covalent-irreversible. Allosteric effectors combine with enzymes in a noncovalent, reversible manner to produce a change in the level of enzyme activity. Covalent modification (such as phosphorylation) can result in modulation of enzyme activity. The reversible covalent modification of enzymes is itself enzyme-catalyzed and directed by hormone action. Irreversible proteolytic cleavage can result in conversion of proenzymes to active enzymes. This form of control is best understood for digestive proteases in mammals, but it extends to other types of enzymes as well.

A. ALLOSTERIC ENZYMES

Threonine deaminase catalyzes the first reaction in the sequence by which threonine is converted to isoleucine (Fig. 6.1). The end product of the reaction sequence (L-isoleucine) inhibits threonine deaminase in a very specific way, while the enzyme is unaffected by D-isoleucine and L-valine. Inhibition by L-leucine is 100 times weaker than inhibition by L-isoleucine. This inhibition by an end product (often called feedback inhibition) was viewed initially as a simple mechanism by which a cell could shut down a pathway when a sufficient level of the end product was available. In contrast, the pathway would function with greatest efficiency at very low steady-state levels of the end product. The kinetic results of the inhibition of threonine deaminase were found to be unusual in that L-isoleucine was not exactly competitive, noncompetitive, or uncompetitive. Another very interesting characteristic of threonine deaminase is that, on heating at 55°C, the sensitivity to inhibition by L-isoleucine is lost (4).

It was proposed by Monod, Changeux, and Jacob that L-isoleucine was inter-

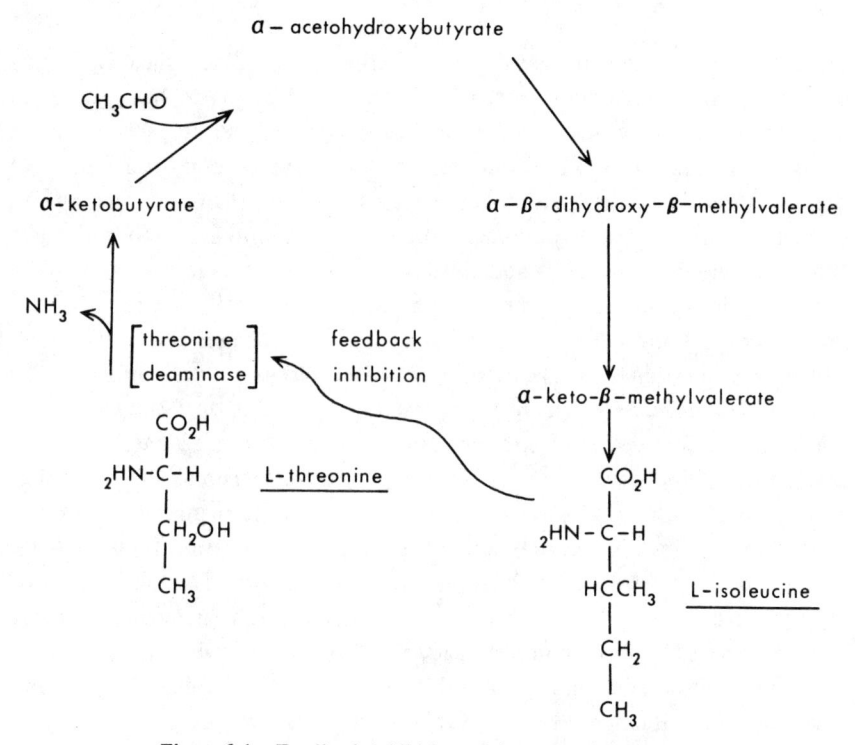

Figure 6.1. Feedback inhibition of threonine deaminase.

acting with a site on the threonine deaminase molecule that was structurally distinct and removed from the active center. Heating the enzyme destroyed the inhibitor site but left the substrate active center intact. The term "allosteric" effector was coined to describe an inhibitor that bound to a site other than the active center (5). In many cases such inhibitors bear little or no resemblance to the substrate or products of the target enzyme.

Threonine deaminase and many other allosteric enzymes are composed of identical subunits, which means that the allosteric effector and the substrate must bind to the same subunit. There are other enzymes, however, which have nonidentical subunits. Aspartate transcarbamylase (ATCase) is composed of catalytic and regulatory subunits. The catalytic subunits cannot bind effector molecules. Regulatory subunits can bind an end product but have no catalytic activity. Gerhart and Pardee demonstrated desensitization to CTP inhibition by treating ATCase with heat, urea, or thiol reagents. Their results indicated separate sites for substrate and CTP (6). Gerhart and Schachman were able to separate the two types of subunit by sedimentation (7). The large subunits were catalytically active, but unaffected by CTP. The smaller subunits were found to have no catalytic activity, but bound CTP. When two types of subunit were mixed, the sensitivity to inhibition was restored. The use of polyacrylamide gel electrophoresis permitted Weber to establish the molecular weights as 33,000 for the catalytic polypeptide and 17,000 for the regulatory peptide (7a). The catalytic subunit is made up of three polypeptide chains, each with a molecular weight of 33,000. The regulatory subunit is composed of two polypeptide chains. The composition of ATCase is shown in Table 6.1. The complete enzyme is composed of two catalytic subunits and three regulatory subunits, represented by the formula $(C_3)_2 (R_2)_3$, in which C is the catalytic polypeptide and R is the regulatory polypeptide. The spatial arrangement shown in Figure 6.2 is based on electron microscopy (8) and X-ray crystallography (9).

TABLE 6.1. Subunit Composition of ATCase

	Molecular Weight	Number of Polypeptide Chains	Composition
Catalytic subunit	100,000	3	C_3
Component polypeptide chain	33,000	1	C
Regulatory Subunit	34,000	2	R_2
Component polypeptide chain	17,000	1	R
Native enzyme	310,000	12	$(C_3)_2 (R_2)_3$

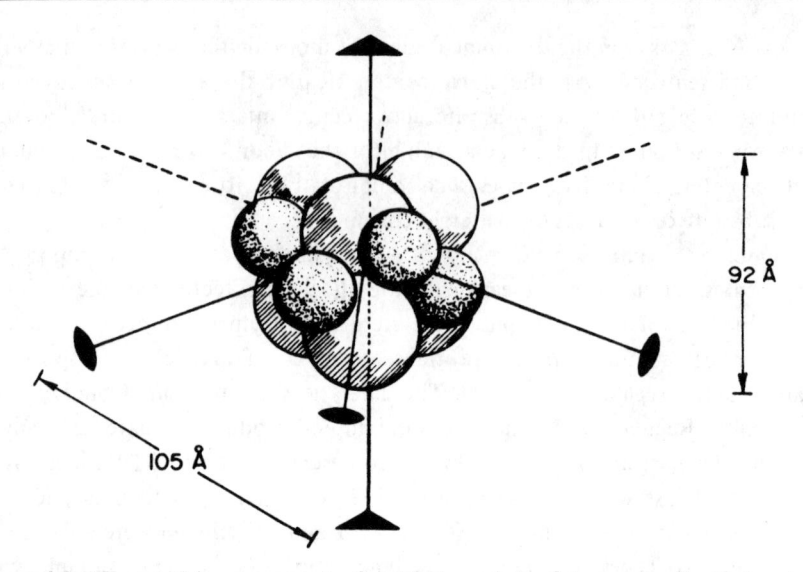

Figure 6.2. The spatial arrangement of catalytic and regulatory subunits of aspartate transcarbamylase.

Allosteric enzymes differ from Michaelis-Menten enzymes in their kinetic behavior. In many known cases allosteric enzymes exhibit *homotropic* activation (Fig. 6.3). The sigmoid curve can be explained by cooperative binding of substrate—that is, binding of the first molecule of substrate to a multisite enzyme facilitates the binding of the second substrate molecule. The s-shaped dependence of rate on S_0 means that the enzyme-catalyzed reaction can go from $v_0/V_m = 0.1$ to $v_0/V_m = 0.75$ with a change in S_0 of 2.3-fold (Fig. 6.4). Such a change for a Michaelis-Menten enzyme requires a change of S_0 of 27-fold.

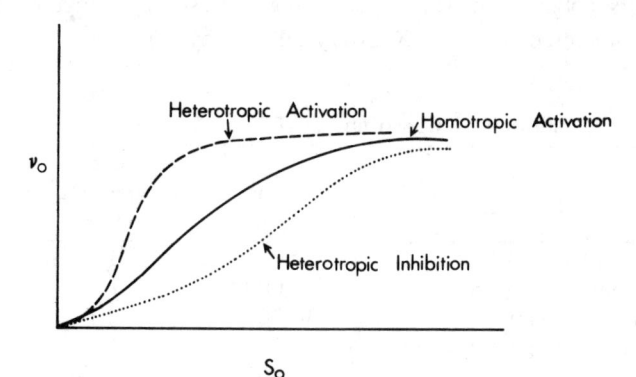

Figure 6.3. Plots of v_0 versus S_0 for an allosteric enzyme.

A. ALLOSTERIC ENZYMES

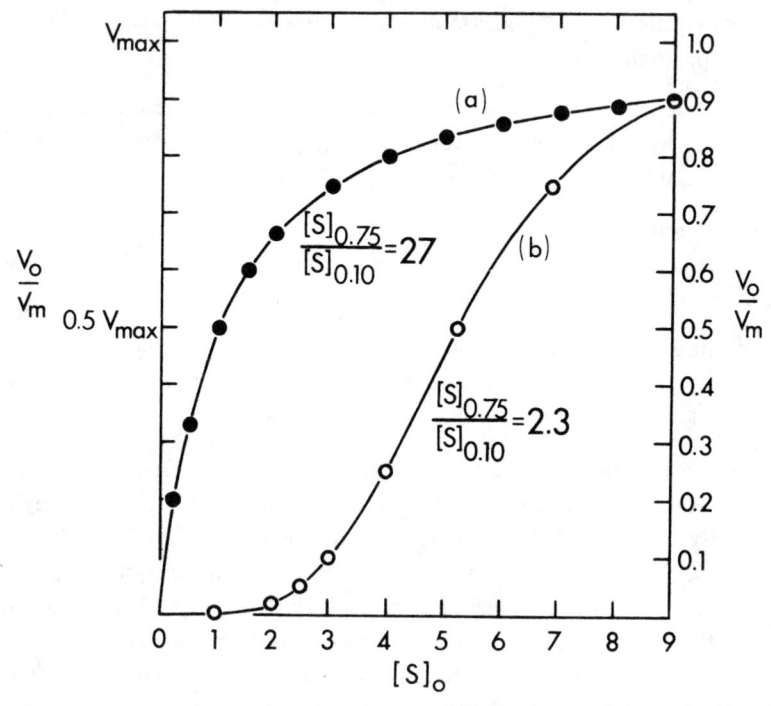

Figure 6.4. Comparison of velocity curves for two different enzymes that, coincidentally, have the same v_0 at [S] = 9. (*a*) Hyperbolic response. (*b*) Sigmoidal response.

Heterotropic effectors are molecules other than the substrate. They can be positive or negative in their action. A heterotropic activator moves the sigmoid curve to the left (Fig. 6.3). The enzyme-catalyzed rate will reach the point of half-saturation ($v_0/v_m = 0.5$) with a lower substrate concentration (S_5) than when activator is absent. A heterotropic inhibitor moves the sigmoid curve to the right with the opposite effect. A greater level of S_0 must be provided to attain the point of half-saturation.

These effects were first explained quantitatively by Monod, Wyman, and Changeux in their classic paper of 1965 (10). Their MWC, or symmetry, model has six basic features.

1. Allosteric proteins are oligomers composed of identical monomers (protomers) that are associated in such a way that they all occupy equivalent positions.

2. Only one site on each protomer corresponds to each ligand able to form a stereospecific complex with the protein.

3. The conformation of each protomer is constrained by its association with the other protomers.

4. At least two states are reversibly accessible to allosteric oligomers. These states differ by the distribution and/or energy of interprotomer bonds, and so also by the conformational constraints imposed upon the protomers.

5. As a result, the affinity of one (or several) of the stereospecific sites toward the corresponding ligand is altered when transition from one to the other state occurs.

6. When the protein goes from one state to another, its molecular symmetry (including the symmetry of the conformational constraints imposed upon each protomer) is conserved.

The model is illustrated for a dimeric enzyme in Figure 6.5. There are two states, R and T, for which the substrate has different affinities. The substrate may bind very poorly or not at all to the T state. Homotropic activation can be explained by equilibrium arguments. In the absence of substrate the enzyme is mainly in the T state. The conformational equilibrium is shifted to the left when substrate binds to the R form. Since the oligomer has multiple binding sites, the binding of the first molecule results in the production of additional binding sites. The result is cooperative binding.

Using methods derived from those employed by Klotz for noncooperative systems (11), Monod, Wyman, and Changeux derived equations to explain coopera-

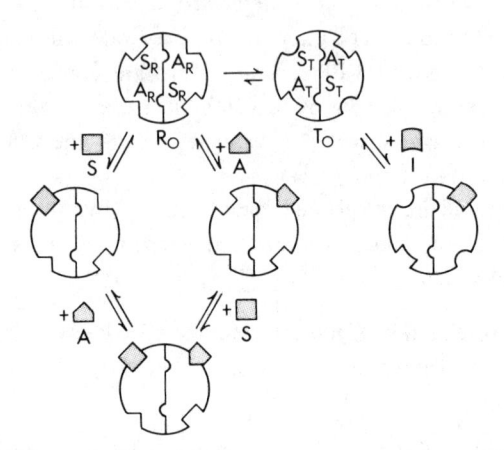

Figure 6.5. The MWC, or symmetry, model for allosteric enzymes.

tive substrate binding. For a dimeric system in which the substrate cannot bind to the T state, the following equations may be written:

$$R_0 \rightleftharpoons T_0; \quad L = T_0/R_0 \tag{1}$$

$$R_0 + S \rightleftharpoons R_1 \tag{2}$$

$$R_1 + S \rightleftharpoons R_2 \tag{3}$$

$$K_R = \frac{2[R_0][S]}{[R_1]} = \frac{[R_1][S]}{2[R_2]} \tag{4}$$

K_R represents the microscopic dissociation; 2's in Eq. 4 are statistical corrections (11):

$$K_R/2 = \frac{[R_0][S]}{R_1} = K_1 \tag{5}$$

$$2K_R = \frac{[R_1][S]}{[R_2]} = K_2 \tag{6}$$

The fraction of saturation is given by Y (10):

$$Y = \frac{[\text{occupied sites}]}{[\text{total sites}]} = \frac{[S]/K_R\,(1 + [S]/K_R)}{L + (1 + [S]/K_R)^2} \tag{7}$$

A plot of Y versus $[S]$ is sigmoidal. Since

$$\text{rate} = YV_{max} \tag{8}$$

it follows that a plot of rate versus $[S]$ will also be sigmoidal. The greatest cooperativity arises when substrate binds only to R forms and when L is large. What would happen to Eqs. 7 and 8 if L were zero?

An empirical measure of cooperativity is given by the Hill coefficient. Consider the ideal case, in which n molecules of substrate leap onto the enzyme simultaneously.

$$E + nS \rightleftharpoons ES_n \tag{9}$$

(Even when n equals 2, this is most unlikely.) For the idealized case,

$$K = [E][S]^n/[ES_n] \tag{10}$$

The extent of saturation is given by Y

$$Y = [ES_n]/E_0 \tag{11}$$

$$1 - Y = [E]/E_0 \tag{12}$$

Combination of Eqs. 10, 11, and 12 gives

$$\log[Y/(1-Y)] = n\log[S] - \log K \tag{13}$$

$$\log[v/V_m - v] = n\log[S] - \log K \tag{14}$$

A plot of $\log[v/V_m - v]$ versus log S is known as a Hill plot. The slope, n, is the Hill coefficient. For a Michaelis-Menten enzyme, $n = 1$. For an allosteric enzyme with four protomers, n may approach 3, which is an indication of strongly cooperative binding.

The qualitative explanation for heterotropic effects (Fig. 6.3) is similar to that for homotropic activation. When an activator binds to the R form, the equilibrium of R and T shifts to the left, which results in more binding sites for substrate (Fig. 6.5). Conversely, when an inhibitor binds to T, the equilibrium shifts to the right, which results in fewer substrate-binding sites.

Homotropic inhibition has been observed in some instances, and here binding of the first molecule of substrate is *stronger* than binding of the second molecule. For such enzymes Hill coefficients are less than one. Enzyme systems with Hill coefficients less than one are said to exhibit negative cooperativity (12). The term *anticooperative* is used by some. Conway and Koshland first observed negative cooperativity with glyceraldehyde-3-phosphate dehydrogenase. The binding of four molecules of NAD$^+$ was described by the following binding constants: $K_1 < 10^{-11}$ M, $K_2 < 10^{-9}$ M, $K_3 = 3 \times 10^{-7}$ M, and $K_4 = 2.6 \times 10^{-5}$ M. As we shall see below, negative cooperativity cannot be explained by the MWC model.

A phenomenon that may be considered a special case of negative cooperativity is known as half-of-the-sites reactivity. An enzyme having four sites can react with only two substrate molecules or labeling reagents. Most enzymes that show half-site reactivity do so with nonspecific substrates or with site-directed reagents. Tyrosyl-t-RNA synthetase, which has two sites, exhibits half-site reac-

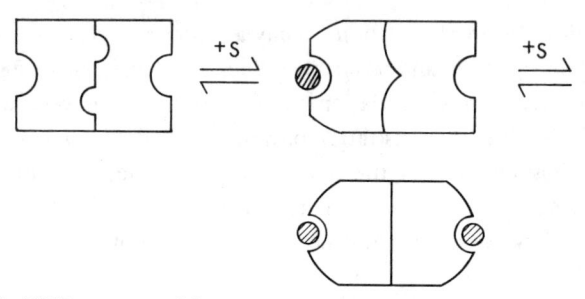

Figure 6.6. The KNE, or sequential, model, showing conformational change in the subunits of an allosteric enzyme.

tivity with specific substrates. The second site reacts, but at a rate 10^4 times slower than the first (14). Half-site reactivity appears to be in violation of nature's law of molecular economy: why make a molecule with four identical protomers and four active sites if only two will be used? Although some half-site reactivity can be explained by nonhomogeneous enzyme preparations or other artifacts, as Fersht suggests (15), there are bona fide examples, such as tryosyl-*t*-RNA synthetase. A complete understanding of the chemical and biological basis of half-site reactivity is not yet available.

Negative cooperativity cannot be explained by the MWC model (Fig. 6.5). Notice that in this symmetry model the nature of the sites does not change. Since substrate changes only the distribution of sites, a reduction in subsequent binding affinity when a site is occupied by substrate is not possible. A model to explain both positive and negative cooperativity was put forth by Koshland, Nemethy, and Filmer (16). This model (the KNF, or sequential, model) involves a conformational change of the subunit to which the ligand or substrate binds. The adjacent subunit is affected through the subunit contact region. Binding of a second molecule of substrate may be either stronger (positive cooperativity) or weaker (negative cooperativity) than the first. In the KNF model, conformational changes of subunits are sequential. In the MWC model, the changes occur in a concerted or all-or-none manner. These points are illustrated in Figures 6.5 and 6.6.

B. CONTROL BY ENZYME-CATALYZED CHEMICAL MODIFICATION (REVERSIBLE)

Enzymes can be activated and deactivated by reactions catalyzed by other enzymes. The activating (deactivating) enzymes can, in turn, be controlled by hormones. Chemical modifications known to be involved with regulation include phosphorylation/dephosphorylation, adenylation/deadenylation, ADP ribosylation/de-ADP ribosylation, and reversible reduction of disulfide bridges (17).

Enzyme-catalyzed phosphorylation/dephosphorylation is very widespread in biology (18). The first demonstration of covalent modulation was the activation of glycogen phosphorylase by the transfer of a phosphoryl group from ATP to the hydroxyl group of a specific seryl residue (19; Fig. 6.7). Phosphorylase catalyzes the breakdown of glycogen, which is the storage polysaccharide in animals:

$$(\text{Glucose})_n + \text{Pi} \rightarrow (\text{Glucose})_{n-1} + \text{glucose 1 - phosphate}$$

In the presence of phosphorylase kinase, phosphorylase b (the less active form) is transformed to phosphorylase a, the more active form. Removal of the phosphate groups catalyzed by phosphorylase phosphatase results in inactivation of the enzyme (Fig. 6.7).

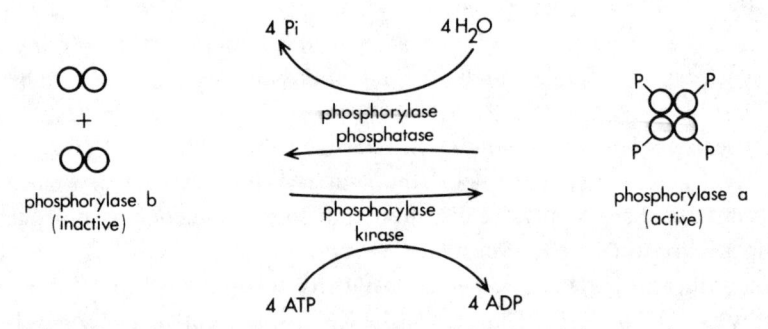

Figure 6.7. The enzyme-catalyzed covalent modulation of glycogen phosphorylase.

Krebs and coworkers showed that phosphorylase kinase was also activated by an enzyme-catalyzed phosphorylation (19). The conversion is stimulated by cyclic AMP.

The enzyme is called phosphorylase kinase kinase, or cyclic AMP-dependent protein kinase (cAMP Prk). The mode of action of cAMP is that it disrupts the association of regulatory subunits with catalytic subunits, as shown below (20).

$$R_2C_2 + 2\,\text{cAMP} \rightarrow (\text{cAMP})_2 R_2 + 2\,C$$

(inactive) (active)

Epinephrine (adrenalin) activates adenylate cyclase, which is the enzyme responsible for the production of cAMP from ATP:

B. CONTROL BY ENZYME-CATALYZED CHEMICAL MODIFICATION 171

$$\text{ATP} \xrightarrow{\text{adenylate cyclase}} \text{cAMP} + \text{PPi}$$

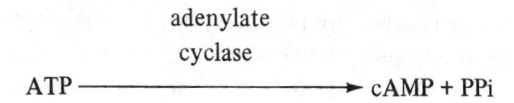

The total picture, which shows epinephrine stimulation of glycogen breakdown in the liver, appears in Figure 6.8. Stimulus, such as fear or anger, to the adrenal medulla results in secretion of epinephrine into the blood. Epinephrine binds to receptors on the outer surface of liver cells, and a conformational change results in the activation of adenylate cyclase; cAMP is produced, which brings about activation of cAMP Prk. Phosphorylase kinase is then activated. The final activation step is the phosphorylation of phosphorylase b, which yields active phosphorylase a. A fundamental feature of the control of glycogen phosphorylase activity is amplification. A very small release of epinephrine results in a rapid increase in blood glucose because of enzyme activation steps. An ampli-

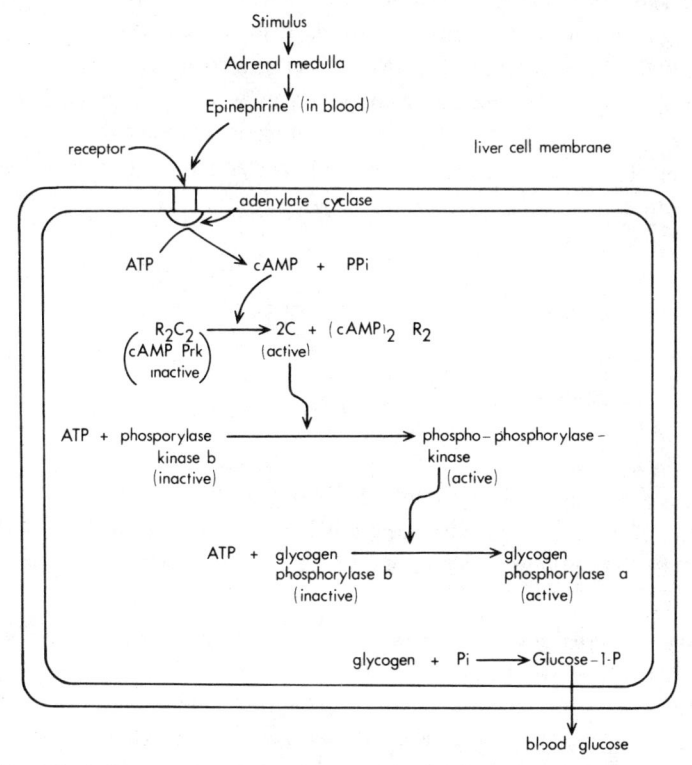

Figure 6.8. The influence of the hormone epinephrine on covalent modulation of glycogen phosphorylase.

fication effect occurs because the production of one molecule of active enzyme gives rise to many molecules of the product of the reaction catalyzed by that particular enzyme. Schemes such as Figure 6.8, which embody multiple amplification stages, have been termed "amplification cascades." The presence of a few molecules of hormone at the "top" of the process brings about a veritable avalanche of product at the bottom.

Another example of enzyme-catalyzed chemical modification involves the glutamine synthetase of *E. coli*. This enzyme catalyzes the formation of glutamine from glutamic acid, ATP, and ammonia. Stadtman, Holzer, and others have found that the glutamine synthetase is involved in a complex regulatory system that embodies feedback inhibition by many effectors and also a covalent modification-adenylation/deadenylation (17, 21).

In Holzer's laboratory in 1966 Mecke and others discovered "glutamine synthetase-activating enzyme", which rapidly deactivates glutamine synthetase (22). In 1967 Stadtman and his coworkers found that glutamine synthetase from *E. coli* grown under different conditions had varying amounts of covalently bound AMP (23). The following year Holzer and coworkers showed that his inactivating enzyme and the adenylating enzyme were the same (24). Glutamine synthetase of *E. coli* is composed of 12 subunits—two hexamers, one on top of the other. The adenyl transferase (ATase) catalyzes the attachment of AMP to all 12 protomers:

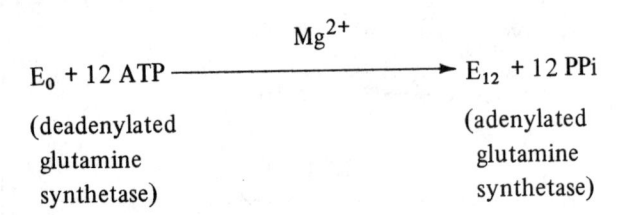

$$E_0 + 12\,ATP \xrightarrow{Mg^{2+}} E_{12} + 12\,PPi$$

(deadenylated glutamine synthetase)   (adenylated glutamine synthetase)

Synthetase activity of E_0 requires Mg^{2+}. The adenylation results in an enzyme form, which is Mn^{2+}-dependent and much less active than the unadenylated enzyme. The pH optimum is also changed, along with sensitivity to some feedback inhibitors.

Holzer and collaborators showed that the deactivated enzyme could be revived by transferring it from a medium rich in NH_4^+ to one containing low levels of NH_4^+ (25). This finding suggested that the adenylation can be reversed and that the deadenylation is subject to control. It was found by Shapiro that the adenylation/deadenylation system has more than one component (26).

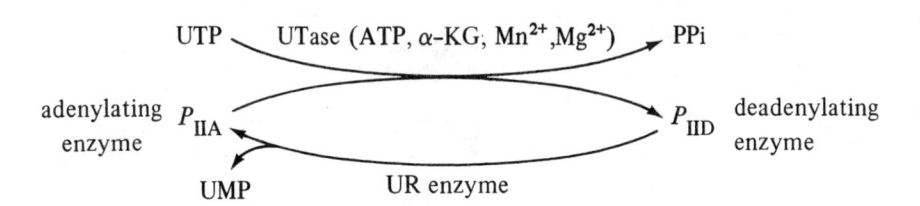

B. CONTROL BY ENZYME-CATALYZED CHEMICAL MODIFICATION

The enzyme system associated with deadenylation was resolved into two fractions, P_I and P_{II}. In addition to P_I and P_{II}, the deadenylation reaction required α-ketoglutarate, UTP, ATP, and P_i. Inorganic phosphate is a reactant, since deadenylation is a phosphorolysis reaction:

$$E_n-\left(\langle O \rangle -O-AMP\right) \xrightarrow[nPi \quad nATP]{\text{deadenylation system}} E_0-\left(\langle O \rangle -OH\right)$$

Brown and coworkers showed that P_{II} could exist in two forms—P_{IIA} and P_{IID}, one for adenylation and one for deadenylation (27). UTP was found to react with P_{IID} to form a uridylyl enzyme. Hence, the glutamine synthetase system is a second example of covalent modification of an enzyme responsible for catalyzing covalent modulation of another enzyme. The attachment and detachment of the uridylyl group are catalyzed by components of the P_I fraction (28).

$$\text{adenylating enzyme } P_{IIA} \xrightleftharpoons[\substack{UMP \\ \text{UR enzyme}}]{\substack{UTP \quad \text{UTase (ATP, α-KG; Mn}^{2+}\text{,Mg}^{2+}\text{)} \quad PPi}} P_{IID} \text{ deadenylating enzyme}$$

Brown and coworkers showed that P_{II} could exist in two forms—P_{IIA} and P_{IID}, one for adenylation and one for deadenylation (27). UTP was found to react with P_{IID} to form a uridylyl enzyme. Hence, the glutamine synthetase system is a second example of covalent modification of an enzyme responsible for catalyzing covalent modulation of another enzyme. The attachment and detachment of the uridylyl group are catalyzed by components of the P_I fraction (28).

UTase, uridylyl transferase, is activated by α-ketoglutarate, ATP, Mg^{2+}, and Mn^{2+}. The uridylyl-removing enzyme (UR enzyme) in the presence of Mg^{2+} is activated by either α-ketoglutarate or ATP. This finding (*in vitro*) appears to be inconsistent with the observation that α-ketoglutarate and ATP also activate the UTase. One possible explanation is that, *in vivo*, the enzyme is activated by

TABLE 6.2. Enzymes and Activators Involved in the Control of Glutamine Synthetase by Covalent Modification

Activity	Effectors Activators	Inhibitors
Adenylylation (ATase-P_{IIA})	Glutamine + Me^{2+}	α-Ketoglutarate
Deadenylylation (ATase-P_{IID})	α-Ketoglutarate + ATP + Me^{2+}	Glutamine
Uridylylation (UTase)	α-Ketoglutarate + ATP + Me^{2+}	Glutamine
Deuridylylation (UR-enzyme)	Mn^{2+} (or Mg^{2+} + α-KG + ATP)	

Mn^{2+}. The manganese-activated enzyme does not require α-ketoglutarate or ATP. The activators for the glutamine synthetase system appear in Table 6.2.

It is clear that glutamine synthetase can be activated in cascade fashion. ATP and α-ketoglutarate can activate UTase, which catalyzes uridylylation of P_{IIA} to give P_{IID}, the deadenylating enzyme. P_{IID} can then activate many glutamine synthetase molecules by catalyzing the removal of AMP. This sequence is illustrated in Figure 6.9.

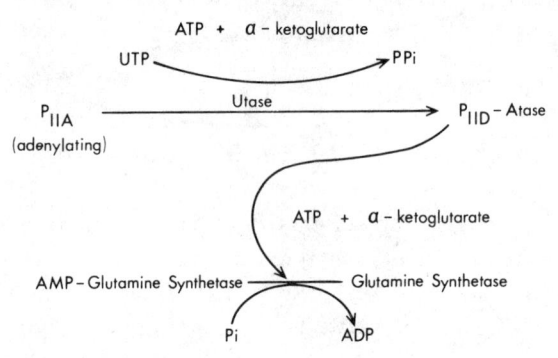

Figure 6.9. The cascade leading to activated glutamine synthetase.

C. CONTROL BY PROTEOLYTIC ACTION (30)

Activation of Pancreatic Zymogens (20)

Ingested protein must be broken down into free amino acids and small peptides before absorption in the small intestine is possible. The fragmentation of dietary protein is accomplished by hydrolysis reactions, which are catalyzed by proteolytic enzymes synthesized in the stomach wall and the acinar cells of the exocrine pancreas. An important question asked many years ago is, What prevents the digestion of the tissue at the site of synthesis? It was found that pancreatic juice was not active until it entered the small intestine. The factor that brings about the activation was termed "enterokinase." This enzyme (which has nothing to do with phosphoryl group transfer) is made in the brush border cells of the small intestine. Enterokinase catalyzes the cleavage of the bond shown below.

$$\text{H-Val-(Asp)}_4 \text{ Lys-Ile-Val—bovine trypsinogen}$$
$$15 \uparrow 16$$
$$\text{Enterokinase cleavage}$$

The removal of the hexapeptide creates a new N-terminus. Asp-194 and Ile-16 (chymotrypsinogen numbering system) form a salt bridge with a concomitant conformational change, which activates the enzyme.

The activation of trypsinogen is autocatalytic. Trypsin produced initially by enterokinase will activate other molecules of trypsinogen. Trypsin, in turn, activates the other proteases shown in Table 6.3. Procarboxypeptidase A is activated by chymotrypsin C. The activation of pancreatic zymogens is another example of the amplification cascade in regulation of enzyme action.

It would be disastrous if the pancreatic proteases were activated prematurely. To prevent this, an inhibitor of trypsin is present. This inhibitor, a protein with a molecular weight of 6500, binds very tightly to trypsin ($K_D = 10^{-12}$ M). About two percent of the protein in pancreatic juice is trypsin inhibitor, an amount adequate to tie up any active trypsin in the pancreatic juice. But it is quickly overwhelmed in the small intestine, where enterokinase is available to convert trypsinogen.

Proteolytic Activation and Blood Clotting

Coagulation of blood is a complex biochemical process involving many proteolytic activations (31); multiple zymogen activation steps in sequence provide a high degree of amplification. A simplified version of the coagulation process is shown in Figure 6.10. There are two pathways for initiation. The intrinsic pathway is initiated by blood contact with an unnatural surface; the factors involved are normally found in blood plasma. The extrinsic pathway involves activation by factors present in normal plasma and also in tissues. Hence, this pathway can be stimulated by factors released by traumatized tissue. The pathways converge with the production of proteases, which activate Factor X to Factor Xa. Pro-

TABLE 6.3. Proteolytic Enzymes and Their Zymogens

Zymogen ⟶ Active Enzyme	Site of Zymogen Synthesis	Type
Pepsinogen $\xrightarrow{H^+}$ pepsin	Stomach	Endopeptidase
Chymotrypsinogens A, B, and C $\xrightarrow{trypsin}$ chymotrypsins A, B, and C	Pancreas	Endopeptidase
Proelastase $\xrightarrow{trypsin}$ elastase	Pancreas	Endopeptidase
Procarboxypeptidase B $\xrightarrow{trypsin}$ CPB	Pancreas	Exopeptidase
Procarboxypeptidase A $\xrightarrow{chymotrypsin\ C}$ CPA	Pancreas	Exopeptidase
Trypsinogen $\xrightarrow{enterokinase}$ trypsin	Pancreas	Endopeptidase

C. CONTROL BY PROTEOLYTIC ACTION (30)

thrombin is activated by Factor Xa by means of two cleavages, involving first an Arg–Thr bond and then an Arg–Ile bond. Thrombin (molecular weight, 33,700) is composed of two chains. One chain bears a strong homology in amino acid sequence to trypsin, chymotrypsin, and elastase. It resembles trypsin in specificity. The role of thrombin is to cleave four Arg–Gly bonds in fibrinogen. Release of these peptides produces fibrin monomers, which then aggregate. A stable clot is formed when a transamidase catalyzes crosslink formation between lysine and glutamine with the liberation of ammonia:

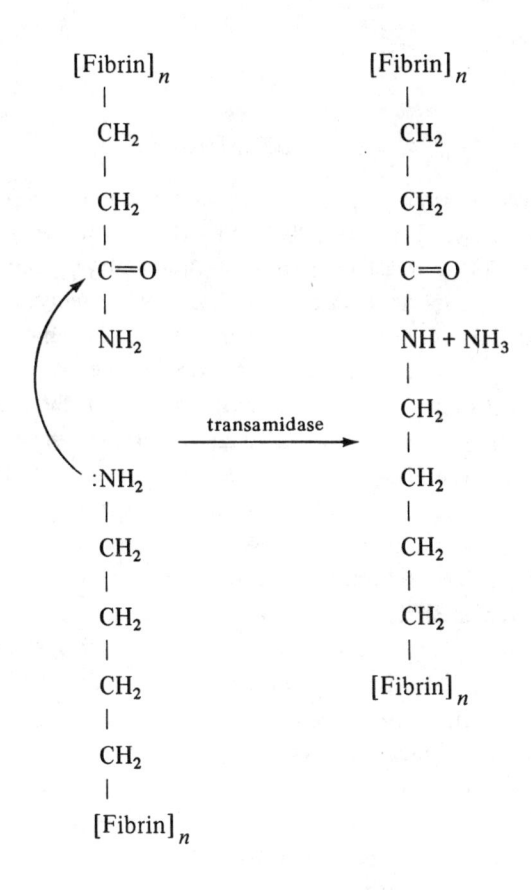

$$
\begin{array}{c}
[\text{Fibrin}]_n \\
| \\
CH_2 \\
| \\
CH_2 \\
| \\
C=O \\
| \\
NH_2 \\
\\
:NH_2 \\
| \\
CH_2 \\
| \\
CH_2 \\
| \\
CH_2 \\
| \\
CH_2 \\
| \\
[\text{Fibrin}]_n
\end{array}
\quad \xrightarrow{\text{transamidase}} \quad
\begin{array}{c}
[\text{Fibrin}]_n \\
| \\
CH_2 \\
| \\
CH_2 \\
| \\
C=O \\
| \\
NH + NH_3 \\
| \\
CH_2 \\
| \\
CH_2 \\
| \\
CH_2 \\
| \\
CH_2 \\
| \\
[\text{Fibrin}]_n
\end{array}
$$

```
Intrinsic pathway:          Extrinsic pathway:
initiated by contact        initiated by trauma
with abnormal surface       to tissue
                    ↓    ↓
                    ↓    ↓
              Factor X ──→ Factor Xa
              prothrombin ──→ thrombin
              fibrinogen ──→ fibrin (clot)
                              transamidase
              stable blood clot
```

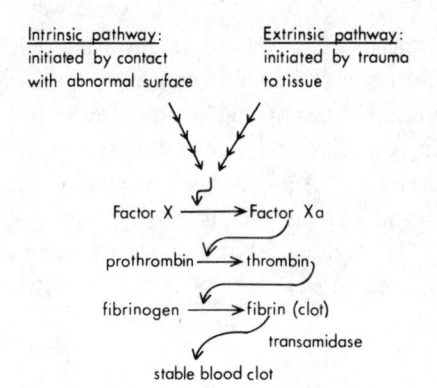

Figure 6.10. An abbreviated scheme for blood clotting.

D. CONCLUSION

Proteolytic activation is common in biological regulation. Two examples, in addition to those discussed here, are activation of chitin synthase, which is involved in septum formation in yeast (32), and activation of the complement system (33). There is an important distinction between proteolytic activation and enzyme-catalyzed interconversion of enzymes by covalent modification. In the activation of the pancreatic zymogens, for example, when the zymogens contact enterokinase, an irreversible avalanche is set off. Indeed, the proteases that are formed digest themselves and dietary protein as well; the pancreatic enzymes are themselves broken down and absorbed. This phenomenon is an economy measure as much as anything, since 100 g of proteases per day are secreted by the pancreas. This amount, in many cases, exceeds dietary protein intake. Interconversion of enzymes by chemical modification is reversible. Also, this mechanism provides for incremental response to a change in metabolic conditions. Another important feature of enzyme interconversion is that such systems are often dependent on noncovalent allosteric effectors. Hence, fine-tuned response in a positive or negative direction is possible.

Another important element in enzyme regulation is physical, chemical association of enzymes with proteins and membranes within the cell. In yeast, there are three well-characterized proteases, each of which is associated with a specific protein inhibitor. These inhibitors restrict the protease activity except when needed at a specific time and place.

Finally, the location of an enzyme in the cell has definite influence on catalytic activity [(35); see also Chapter 7, p. 181]. Complex mechanistic schemes elaborated on the basis of *in vitro* studies may be of questionable physiological significance. Changes in the microenvironments of enzymes within the cytoplasm, membranes, or multienzyme complexes are obvious points for metabolic control (34). It is understandable that little information is available on enzyme location, translocation, and microenvironment. To study is to perturb, which means that the biological uncertainty principle applies and that inference will provide the best possible picture.

REFERENCES

1. *Molecular Mechanisms of Protein Biosynthesis* (H. Weissbach and S. Pestka, Eds.), Academic Press, New York (1977).
2. N. Katunuma, E. Kominami, Y. Banno, K. Kito, Y. Aoki, and G. Urata, in *Advances in Enzyme Regulation* Vol. 14 (G. Weber, Ed.), Pergamon Press, Oxford (1976), p. 325.
3. H. E. Umbarger, *Ann. Rev. Biochem.* 38, 323 (1969).
4. J. P. Changeux, *Cold Spring Harbor Symp. Quant. Biol.* 26, 313 (1963).
5. J. Monod, J. P. Changeux, and F. Jacob, *J. Mol. Biol.* 6, 306 (1969).
6. J. C. Gerhart and A. B. Pardee, *J. Biol. Chem.* 237, 891 (1962).
7. J. C. Gerhart and H. K. Schachman, *Biochemistry* 4, 1054 (1965).
7a. K. Weber, *Nature* 218, 1116 (1969).
8. J. A. Cohlberg, V. P. Pigiet, Jr., and H. K. Schachman, *Biochemistry* 11, 3396 (1972); K. E. Richards and R. C. Williams, *Biochemistry* 11, 3393 (1972).
9. S. G. Warren, B. F. P. Edwards, D. R. Evans, D. C. Wiley, and W. N. Lipscomb, *Proc. Natl. Acad. Sci. USA* 70, 117 (1973).
9a. E. R. Kantrowitz, S. C. Pastra-Landis, and W. N. Lipscomb, *TIBS* 5, 124 (1980).
10. J. Monod, J. Wyman, and J. P. Changeux, *J. Mol. Biol.* 12, 88 (1965).
11. I. M. Klotz, in *The Proteins*, Vol. 1 (H. Neurath and K. Bailey, Eds.), Academic Press, New York (1953), p. 773.
12. A Conway and D. E. Koshland, Jr., *Biochemistry* 7, 4011 (1968).
13. F. Seydoux, O. P. Malhotra, and S. A. Bernhard, *Crit. Rev. Biochem.* 2, 227 (1974).
14. A. R. Fersht, R. S. Mulvey, and G. L. E. Koch, *Biochemistry* 14, 13 (1975).
15. A. R. Fersht, *Enzyme Structure and Mechanism*, Freeman, San Francisco (1977), p. 215.
16. D. E. Koshland, Jr., G. Nemethy, and D. Filmer, *Biochemistry* 5, 365 (1966); D. E. Koshland, Jr., *The Enzymes* Vol. 1, 3rd ed. (P. D. Boyer, Ed.), Academic Press, New York (1970), p. 341.
17. H. Holzer and W. Duntze, *Ann. Rev. Biochem.* 40, 345 (1971).
18. C. S. Rubin and O. M. Rosen, *Ann. Rev. Biochem.* 44, 831 (1975).
19. E. G. Krebs, *Curr. Top. Cell. Regul.* 5, 99 (1972).
20. P. Cohen, *Control of Enzyme Activity*, Wiley, New York (1976) p. 41 and references.
21. E. R. Stadtman and A. Ginsburg, in *The Enzymes*, Vol. 10, 3rd ed. (P. D. Boyer, Ed.),

Academic Press, New York, (1974), p. 755; E. R. Stadtman and P. B. Chock, *Curr. Top. Cell. Regul.* **13**, 53 (1978).
22. D. Mecke, K. Wuff, K. Liess, and H. Holzer, *Biochem. Biophys. Res. Commun.* **24**, 452 (1966).
23. B. M. Shapiro, H. S. Kingden, and E. R. Stadtman, *Proc. Natl. Acad. Sci. USA* **58**, 1703 (1967).
24. H. Holzer, H. Schutt, Z. Masek, and D. Mecke, *Proc. Natl. Acad. Sci. USA* **60**, 721 (1968).
25. H. Holzer, D. Mecke, K. Wulff, K. Liess, and L. Heilmeyer, *Adv. Enzyme Reg.* **5**, 211 (1967).
26. B. M. Shapiro, *Biochemistry* **8**, 659 (1969).
27. M. S. Brown, A. Segal, and E. R. Stadtman, *Proc. Natl. Acad. Sci. USA* **68**, 2949 (1971).
28. J. H. Mangum, G. Magni, and E. R. Stadtman, *Arch. Biochem. Biophys.* **158**, 514 (1973).
29. S. P. Adler, D. Purich, and E. R. Stadtman, *J. Biol. Chem.* **250**, 6264 (1975).
30. H. Holzer and P. C. Heinrich, *Ann. Rev. Biochem.* **49**, 61 (1980).
31. E. W. Davie and K. Fujikawa, *Ann. Rev. Biochem.* **44**, 799 (1975).
32. H. Holzer, in *Advances in Enzyme Regulation*, Vol. 13 (G. Weber, Ed.), Pergamon Press Oxford (1975), p. 132. A description of the work of Lenney, Cabib, and Holzer on control by yeast proteases.
33. H. J. Muller-Eberhard, *Ann. Rev. Biochem.* **43**, 567 (1975).
34. C. J. Masters, *Curr. Top. Cell. Regul.* **12**, 75 (1977).
35. W. Ferdinand, *The Enzyme Molecule*, Wiley, New York (1976), p. 237.

CHAPTER SEVEN

IMMOBILIZED ENZYMES

In cells many enzymes are immobilized; they are fixed in membranes, in multienzyme complexes, or in macrostructures, such as ribosomes. Microenvironmental effects and diffusional limitations are important consequences of immobilization. A membrane-bound enzyme may have altered kinetic properties when removed from its natural location within the cell. As a result of "local" perturbations of $[H^+]$, the pH rate profile of the enzyme in the membrane could be very different from that of the enzyme in solution. The rates of enzyme-catalyzed reactions in solution are rarely controlled by diffusion. In bound enzyme systems, however, diffusional limitations are almost always important. In reaction 1 the rate-limiting step could be diffusion of substrate to the bound enzyme or diffusion of product into the bulk solution.

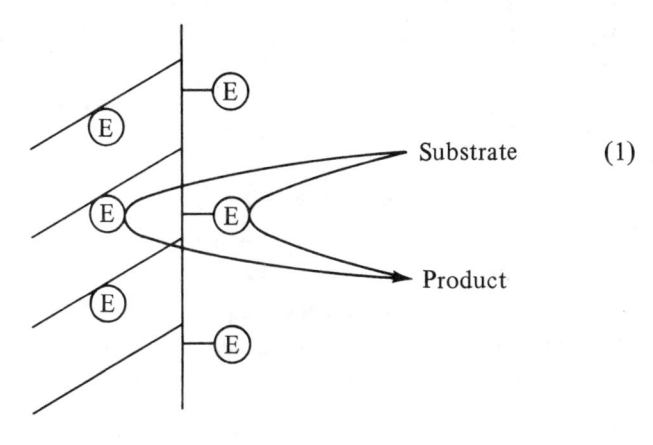

(1)

As we shall see later, the diffusional limitation can be external or internal. External diffusional resistance results from the existence of an unstirred (Nernst) layer on the surface of the particle or structure containing the enzyme. Internal diffusion applies to the transport of substrate to enzyme molecules enbedded within the structure.

The production of soluble enzymes is almost always an expensive process, and much of the work on bound enzymes has centered on practical applications in industry and medicine. Immobilization is economical because it permits repeated use of the enzyme. Enhanced stability is another advantage, especially for systems that involve proteolytic enzymes. Rapid separation of the catalyst from the reaction mixture is also an important practical consideration. Research in this area has led to commercially significant processes in the food and chemical industries.

A. METHODS OF ENZYME IMMOBILIZATION

Physical Methods

There are several approaches and a wide variety of specific methods for rendering enzymes insoluble (Table 7.1). One of the simplest methods is physical adsorption. An example would be the binding of a negatively charged enzyme on a positively charged support such as DEAE-Sephadex.

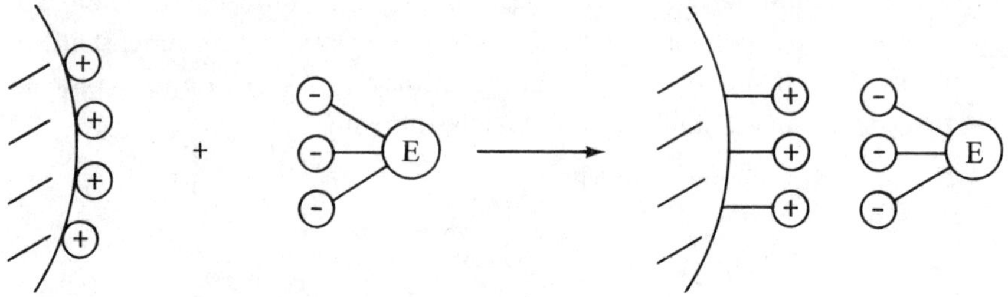

DEAE–Sephadex (2)

At low ionic strength some of these complexes can be quite stable, although gradual leaching occurs. The method is convenient and inexpensive. A commercial process for preparation of optically pure amino acids is based on L-amino acylase bound, according to reaction 2, by Chibata and his associates in Japan (1). Adsorption through apolar bonds is also possible. Hofstee has shown that

A. METHODS OF ENZYME IMMOBILIZATION

TABLE 7.1. Methods of Enzyme Immobilization

A. Physical
1. Adsorption to water-insoluble carrier
2. Entrapment in a matrix by
   a. polymerization of water-soluble monomers in a solution containing enzyme
   b. fiber
3. Microencapsulation—trapping of droplets of enzyme solution using polymerization at a water/organic interface

B. Chemical
1. Covalent attachment of enzymes to the surface of a water-insoluble support
2. Crosslinking a pure enzyme or protein mixture to yield a stable particle or film

---

alkylamino-agarose forms very strong complexes with the following enzymes: xanthine oxidase, lactate dehydrogenase, DNA-ase, alkaline phosphatase, and urease (2). Similar results have been reported for phenoxyacetyl cellulose (3). As expected, the enzymes are released from the apolar supports in the presence of detergents.

Fiber entrapment (4, 5) and microencapsulation (6) are other physical methods for containment of enzymes. Microencapsulation has attracted considerable attention since microcapsules can be viewed as injectable "artificial cells." Nylon microcapsules have been prepared. In one method, an aqueous solution containing the desired enzyme, hemoglobin (as a stabilizer), and hexamethylenediamine is dispersed into an organic solvent containing adipoyl chloride. The reaction of the bifunctional acid chloride and the diamine yields nylon microcapsules with a diameter of about 20 μm.

Entrapment within gels made with water-soluble monomers was introduced in 1963 by Bernfeld and Wan (7) and has been applied to many enzymes since then (8). The technique involves the polymerization of a solution containing a vinyl monomer, a bis-vinyl crosslinking agent, and the enzyme. Initiation is usually accomplished with ammonium persulfate, in a manner analogous to preparation of polyacrylamide gels for electrophoresis. Gels may be prepared in bulk or in beaded form by dispersing the polymerization mixture into an organic solvent.

Chemical Methods

Covalent Attachment to Solid Supports. There have been hundreds of examples of chemical attachment of enzymes to water-insoluble supports. An excellent and thorough review of this subject has been prepared by Goldstein and Manecke (9). The types of support materials that have been used appear in Table 7.2. Since the methods of chemical coupling are so diverse, the types of materials to which enzymes have been bound include such substances as cotton string and polymer-coated brick dust.

Carbohydrates are popular supports for enzyme immobilization and affinity chromatography because of their hydrophilicity and availability. Cellulose, Sephadex, agarose, and so on have been activated by numerous procedures (9, 10). Cyanogen bromide activation of agarose (Sepharose) could be the most frequently used system for immobilization (reaction 3). Agarose is a polymer of β-D-galactose and 3,6-anhydrogalactose.

$$\text{—OH} + \text{CNBr} \longrightarrow \text{—OC}\equiv\text{N} \longrightarrow \text{—OC—NH—}\textcircled{E}$$
$$\textcircled{E}\text{—NH}_2 \qquad \overset{\overset{+}{\text{NH}_2}}{\underset{\|}{}}$$

(3)

The substituted isourea, which is formed primarily by way of the cyanate ester shown in reaction 3, is susceptible to nucleophilic attack. High pH and nucleophilic buffers are not generally used with these enzyme conjugates for extended periods.

Synthetic organic polymers have been employed in a variety of ways for enzyme immobilization. Nylon, phenol-formaldehyde resins, vinyl polymers, and synthetic polypeptides have been used (9).

Inorganic supports are of interest because they can be used in packed beds with high flow rates. Weetall showed that silanized porous glass beads could be used for covalent immobilization of enzymes and antibodies (11). The first step in the activation of porous glass and other materials is the silanization of the surface (Fig. 7.1). The glass beads are reacted with a silanization agent, such as γ-aminopropyltriethoxysilane. The amino group that is introduced can be derivitized in a number of ways. A simple method is the use of a crosslinking reagent, such as glutaraldehyde. Succinylation followed by N-hydroxysuccinimide ester formation is a good activation method.

Inorganic-organic composites have been prepared to obtain a support with the rigidity of glass and the stability of the organic materials. Siliceous materials may be coated easily by treatment with polyethylenediamine in methanol. Crosslinking the polymer layer with glutaraldehyde produces a stable support.

TABLE 7.2. Some Support Materials Classified According to Chemical Composition

Material	Examples
A. Organic	
1. Carbohydrates	Cellulose, agarose, starch
2. Vinyl polymers	Polyacrylamide, poly(ethylene/maleic anhydride), poly(methacrylic acid/methacrylic acid-m-fluoroanilide)
3. Polymers of amino acids and their derivatives, polyamides	Poly(p-amino-D,L-phenyl-alanine/L-leucine), nylon
4. Phenol-formaldehyde	Amine containing resins (e.g., Duolite A-7)
B. Inorganic	
1. Glass	Porous and solids beads
2. Metals	Nickel screen
3. Others	Colloidal silica, aluminas
C. Inorganic-Organic	Conjugates of silicas or porous glass with organic polymers, magnetite polyacrylamide

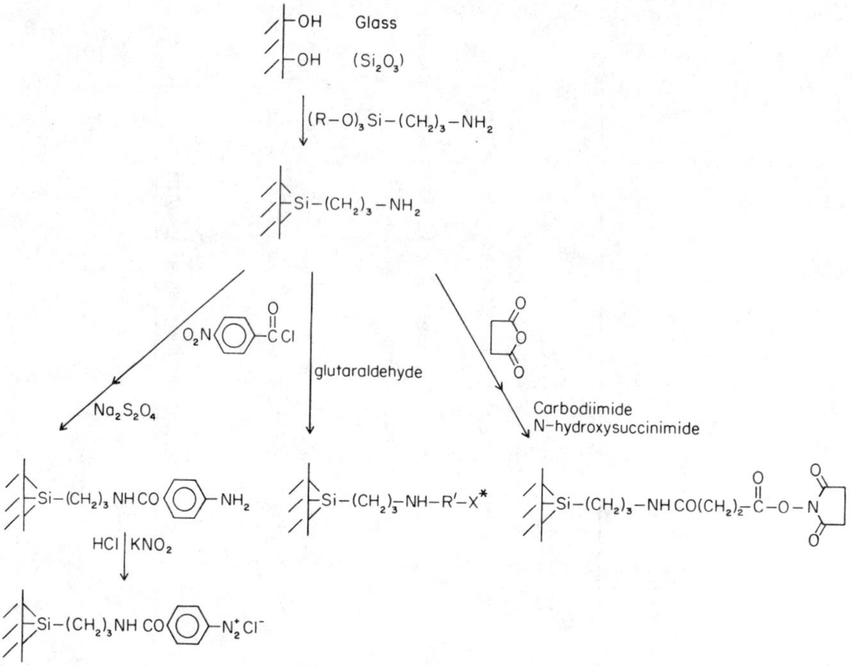

Figure 7.1 Silanization and activation of porous glass.

Konecny and coworkers have used this approach for immobilization of esterases (12). The starting inorganic material for this immobilization is brick dust.

Immobilization by Crosslinking. Enzyme crystals or amorphous solids can be insolubilized by simply soaking them in a solution containing a crosslinking reagent (13). For example, crystals of carboxypeptidase A are insoluble in buffer solutions that have low salt concentrations. After a bifunctional reagent diffuses into the crystal and reacts, the enzyme is insoluble, even in such salt solutions as 1 M NaCl. A compilation of immobilization methods using crosslinking reagents has appeared (9).

B. PROPERTIES OF IMMOBILIZED ENZYMES

Physical and Chemical Modifications

Physical and chemical modification, microenvironmental effects, and diffusional effects are the major considerations when comparing the bound form of an enzyme to the soluble form. The three-dimensional conformation of an enzyme may change when adsorbed to a solid or when incorporated into a biological membrane. The conformational change obviously could perturb the activity. An extreme case is denaturation of proteins at interfaces, especially those involving apolar solvents or materials. Artificial constraints can alter conformational changes related to the allosteric response. We reported evidence that suggested that trypsin bound to porous glass by multiple linkages differed as a consequence of whether a substrate were present during the immobilization process (14).

Direct observation of the conformation of bound enzymes is not easy: interference by the support and low protein concentrations are two problems. A number of attempts have been made, however, Gabel, Steinberg, and Katchalski devised a fluorescence method in which Sephadex- and Sepharose-enzyme conjugates containing high enzyme loadings were used in a reflectance cell. Their results suggested that the enzyme-matrix interaction provided some conformational stability (15). Berliner and collaborators used ESR to study free and bound forms of spin-labeled trypsin (16). Spectra of samples taken before and after immobilization were essentially identical.

An important consequence of the immobilization of enzymes is the stabilization of proteases. Many proteases digest themselves when free in solution. When the enzyme molecules are fixed on the support, autolysis is prevented simply because enzyme-enzyme interaction is not possible. Mixtures of proteases may be stabilized by immobilization, which permits their use in total hydrolysis of polypeptides in preparation for amino acid analysis (17). Enzyme preparations contaminated with a protease would be stabilized by immobilization, since the protease would be fixed or washed away.

The Microenvironment of Fixed Enzymes

Microenvironmental effects on catalysis by bound enzymes may be rationalized in terms of attraction (or repulsion) of substrates, products, protons, inhibitors, and other effectors. Electrostatic effects on K_m and the pH dependence of trypsin bound to a negatively charged polymer were studied by Goldstein, Levin, and Katchalski (18). The carboxylate anions on the matrix perturbed the pH rate profile by more than two pH units to a more alkaline pH value. The local environment could be considered unusually acidic because of the presence of groups attracting protons (Fig. 7.2). This partitioning of protons can be treated rigorously, using approaches that have been applied to ion exchange. The chemical potential of a species in a charged matrix is given by

$$\mu' = \mu^\circ + RT \ln a' + ZF\psi \tag{4}$$

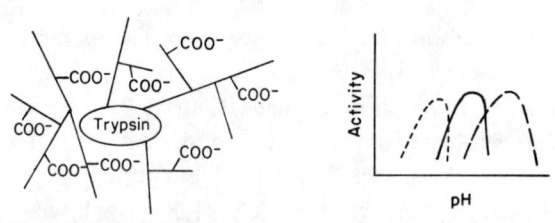

Figure 7.2. (Left) Trypsin immobilized in a negatively charged matrix. (Right) The middle curve is the pH rate profile of a free enzyme. The profile on the more alkaline side is for an enzyme immobilized in a negatively charged matrix. A positively charged matrix shifts the pH optimum to a more acidic pH.

Z is the electrovalency of the species under consideration, F is the Faraday constant, and ψ is the electrostatic potential. For the bulk solution the conventional chemical potential is given by

$$\mu = \mu^\circ + RT \ln a \tag{5}$$

Substraction of Eq. 5 from Eq. 4 gives

$$\mu' - \mu = RT \ln (a'/a) + ZF\psi \tag{6}$$

Division by the Avogadro number and imposing equilibrium ($\mu' - \mu = 0$) yields

$$kT \ln (a'/a) = -Ze\psi$$

or

$$\ln a' - \ln a = Ze\psi/kT \tag{7}$$

in which E is the charge on the electron. When the species under consideration is the proton, Eq. 7 becomes

$$pH' - pH = \Delta pH = 0.43\, e\psi/kT \tag{8}$$

When an enzyme experiences a potential of 100 mv, ΔpH would be 1.7 pH units.
Eq. 7 may be rearranged to

$$a' = a\, e^{-Ze\psi/kT} \tag{9}$$

B. PROPERTIES OF IMMOBILIZED ENZYMES

For a charged substrate

$$[S]' = [S]\, e^{-Ze\psi/kT} \tag{10}$$

The Michaelis-Menten equation is

$$v_0 = \frac{V_m S_0}{K_m + S_0} \tag{11}$$

in which V_m is the maximal velocity, S_0 is the initial substrate concentration, and v_0 is the initial velocity. Combination of Eqs. 10 and 11 gives an expression for the Michaelis constant, K_m', for the bound enzyme system

$$K_m' = K_m\, e^{-Ze\psi/kT} \tag{12}$$

and therefore, for a substrate with a charge of +1

$$\Delta pK_m = pK_m' - pK_m = \log(K_m/K_m') = 0.43\, \psi/kT \tag{13}$$

Goldstein and collaborators determined experimentally the values of ΔpH and ΔpK_m for the ethylene-maleic anhydride/trypsin system (18). The calculated values of ψ (Eqs. 8 and 13) were found to be 92 mv and 96 mv respectively. The fact that these values agree within experimental error supports the theoretical treatment given above. It follows that measured values of ΔpH and ΔpK_m, which accompany removal of an enzyme from a biological (or artificial) structure, are indicative of the environment provided by that structure.

The electrostatic partitioning of protons and substrates would be dependent upon ionic strength. K_m' was found to approach K_m when the ionic strength was raised to 0.5 in the trypsin system (18). Wharton and coworkers derived an equation to relate K_m' to ionic strength:

$$\frac{1}{K_m'} = \frac{I}{\gamma K_m} + \frac{Z_{mc}}{2K_m} \tag{14}$$

in which I is ionic strength, Z_{mc} is the effective concentration of fixed charge groups, and γ is the ratio of mean ion activity coefficients in the matrix and bulk phases (19). Eq. 14 has been verified experimentally and found to be consistent with the results and equations for the trypsin-ethylene-maleic anhydride system (20).

Partitioning of substrate by hydrophobic forces may occur. This type of partitioning has been demonstrated by Brockman and collaborators for a system involving lipase (21). Porous glass was coated with an apolar silane that adsorbs the enzyme lipase. Since the substrate, tripropionine, is also adsorbed, a considerable rate acceleration results. This system is an excellent model of an apolar microenvironment's raising the local concentration of a substrate.

Diffusional Effects

In homogeneous systems the processes of transport of substrate to the enzyme and the formation of the ES complex are rarely rate-limiting. For artificial immobilized enzymes, and almost certainly for naturally occurring bound enzymes, substrate and product transport rates are almost always of kinetic importance. In this discussion basic elements of the subject will be treated. Engasser and Horvath have prepared an excellent comprehensive review of this important subject (22). The interest in diffusional effects is based on the need to optimize immobilized enzyme reactors and to understand the kinetics of fixed enzymes in nature. One important aspect of the latter subject is the kinetic significance of a multienzyme system in which sequential reactions of a metabolic pathway are catalyzed. The following abbreviations will be used:

l	Thickness of the unstirred (Nernst) layer (cm)
D	Diffusion coefficient (cm^2/sec)
V	Volume of the solution (cm^3)
A	Surface area of support material (cm^2)
K_m	Normal Michaelis constant (moles/cm^3)
K'_m	Michaelis constant for bound enzyme (moles/cm^3)
k_{cat}	First-order rate constant for the decomposition of the ES complex to product (sec^{-1})
E'_0	Enzyme loading (moles/cm^2)
E_0	Enzyme concentration (moles/cm^3; $A E'_0/V$)
V'_m	Maximal velocity (moles/cm^2/sec; $k_{cat} E_0'$)
V_m	Maximal velocity (moles/cm^3/sec; $k_{cat} E_0$ or $A V'_m/V$)

B. PROPERTIES OF IMMOBILIZED ENZYMES

S_b Substrate concentration in the bulk solution (moles/cm^3)
S_0 Substrate concentration at support surface (moles/cm^3)
v' Reaction rate (moles/cm^2/sec)
v Reaction rate (moles/cm^3/sec)

External Diffusional Resistance. The unstirred (Nernst) layer that surrounds a particle containing enzyme is shown in Figure 7.3. The rate of substrate diffusion through this layer can be slow in comparison to the enzymatic reactions.

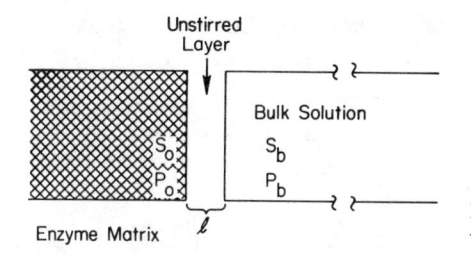

Figure 7.3. The unstirred (Nernst) layer on the surface of an enzyme matrix.

The thickness of the layer, and thus the diffusional resistance, is reduced by increased flow rate or stirring rate. The effects of diffusion in enzyme columns were treated by Hornby and coworkers (23). When the diffusion of substrate through the unstirred layer is rate-limiting, Eq. 15 applies

$$v' = \frac{D}{l}(S_b - S_0) = V_m' \frac{S_0}{(K_m + S_0)} \tag{15}$$

The left side is the diffusion rate and the right side is the enzymatic rate. It can be shown (33, 34) that

$$v' = \frac{V_m' S_b}{K_m' + S_b}$$

or

$$v = \frac{A}{V} v' = \frac{V_m S_b}{K_m' + S_b} \tag{16}$$

Eq. 16 expresses the reaction rate, as determined by product appearance in the bulk phase, as a function of V_m, bulk substrate concentration, and a new K_m– $K_m{}'$–which is expressed by Eq. 17.

$$K_m{}' = K_m + V_m{}' \frac{l}{D} \qquad (17)$$

This equation states that an increase in flow rate through a packed bed would lower K_m as a result of a lower thickness of the unstirred layer, l. The results shown in Figure 7.4 are consistent with Eq. 17.

The electrostatic effects on the external diffusion rate have been studied in several laboratories. The electrostatic term appears in Eq. 15 as

$$v' = \frac{D}{l}(S_b - S_0) - \frac{DZ\,S_b}{RT} F \text{ grad } \psi = \frac{V_m{}' S_0}{K_m{}' + S_0} \qquad (18)$$

In Eq. 18 Z is the electrovalency of the substrate and ψ is the electrostatic potential. In this case $K_m{}'$ is given by

$$K_m{}' = (K_m + V_m{}'\, l/D)\,(l - \lambda)^{-1} \qquad (19)$$
$$\lambda = l\,Z\,F \text{ grad } \psi/RT$$

Internal Diffusional Resistance. Enzyme membranes, gels, and particles heavily loaded with enzyme usually exhibit internal diffusional resistance. This means that the enzyme molecules embedded within the structure are not supplied with substrate. The effectiveness of the bound enzyme is, therefore, less than that of the enzyme in the soluble form. Goldman and coworkers first developed the theory to describe the behavior of the artificial enzyme membranes mentioned above (24). According to Bunting and Laidler (25) the effectiveness factor is given by Eq. 20,

$$v'/v = F = \tanh \gamma x/\gamma x$$
$$\gamma = \left(\frac{k_{\text{cat}} E_0}{4 D K_m}\right)^{1/2} \qquad (20)$$

Eq. 20 was verified by Bunting and Laidler, using thin slices of polyacrylamide gel containing β-galactosidase (25). As the thickness of the gel, x, decreases, the

B. PROPERTIES OF IMMOBILIZED ENZYMES 193

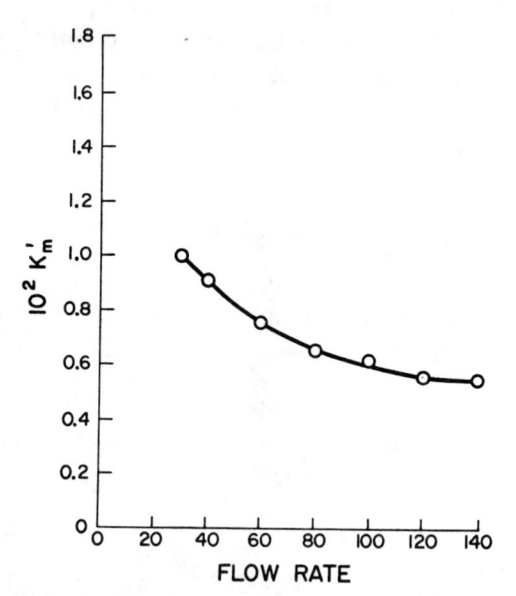

Figure 7.4. The dependence of K_m for an immobilized enzyme on flow rate. The decrease in K_m results from the decrease in the width of the unstirred later (E. 17).

effectiveness factor approaches 1. Notice that k_{cat}/K_m and E_0 are also important. Maximum efficiency was observed for thin slices containing low amounts of enzyme (Fig. 7.5). The experimental points fall very close to the theoretical line.

A related problem in considering the kinetics of bound enzymes is the produc-

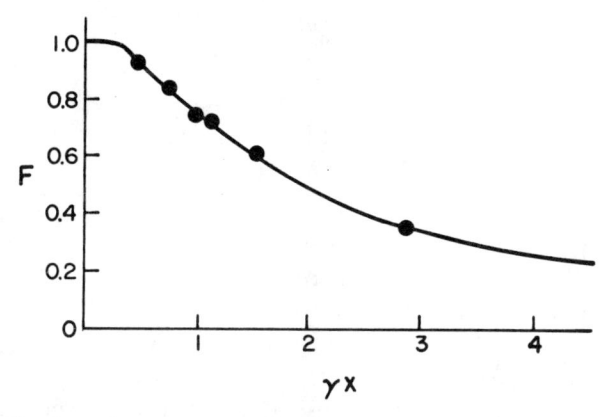

Figure 7.5. The dependence of the effectiveness factor (F) on γ (E. 20) for an artificial gel membrane containing an enzyme.

tion of pH gradients within particles. For enzymes bound to neutral supports, shifts in pH optima have been observed for reactions that result in proton production. These gradients can often be reduced or eliminated by using concentrated buffer solutions.

C. MULTIENZYME SYSTEMS

The study of co-immobilized enzymes (two or more enzymes on the same matrix) is important for modeling the catalysis of sequential reactions by multienzyme complexes and the collections of enzymes in organelles. Consider the scheme below:

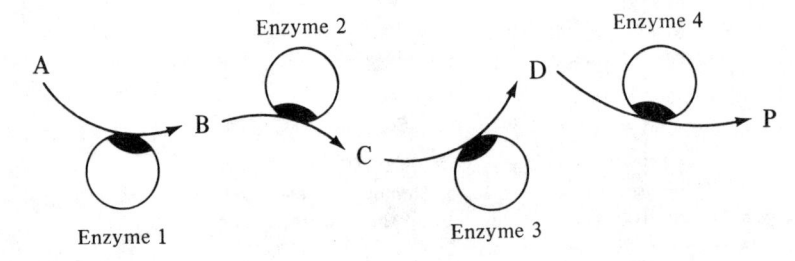

What would be the kinetic difference between free enzyme systems and co-immobilized enzyme systems? Mosbach and Mattiasson addressed this problem by binding hexokinase and glucose-6-phosphate-dehydrogenase to the same matrix (26). The reaction is

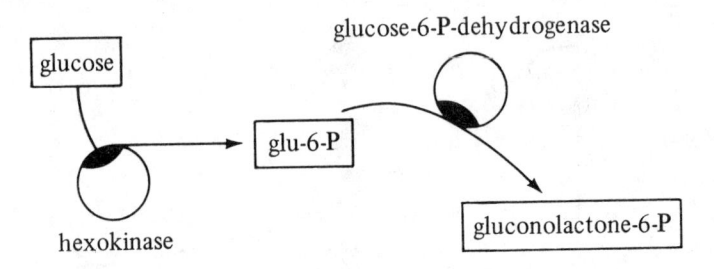

The lag time for the attainment of the steady-state rate was reduced considerably in the case of co-immobilized enzymes, compared to enzymes free in solution. This finding is illustrated in Figure 7.6. The observation may be explained by a buildup of glucose-6-P in the vicinity of the second enzyme, glucose-6-P-dehydrogenase. As expected, a three-enzyme system shows an even greater

C. MULTIENZYME SYSTEMS

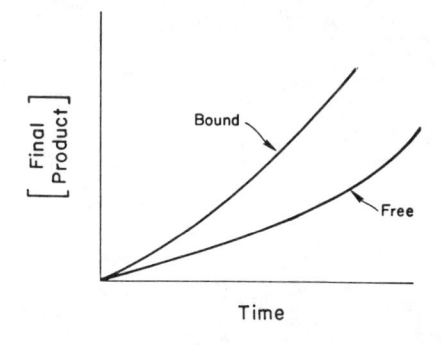

Figure 7.6. Reduction of the lag period for appearance of the final product by co-immobilization of the two enzymes that catalyze sequential reactions.

reduction in lag time (27). In both cases the final steady-state rates of the co-immobilized and soluble systems were identical.

Srere and collaborators (28) immobilized malate dehydrogenase, citrate synthase, and lactate dehydrogenase (Fig. 7.7). The malate-oxalate reaction is thermodynamically unfavorable as written. The co-immobilized enzymes produced a steady-state rate twofold faster than that produced by the soluble enzymes in the absence of pyruvate. In the presence of pyruvate, a rate enhancement of four is observed. These results are explained by an increase in concentration of oxaloacetate in the vicinity of the immobilized citrate synthase, as compared to soluble citrate synthase. The inclusion of pyruvate in the system would tend to increase the $NAD^+/NADH$ ratio, which would in turn increase the amount of oxaloacetate in the microenvironment of citrate synthase. These

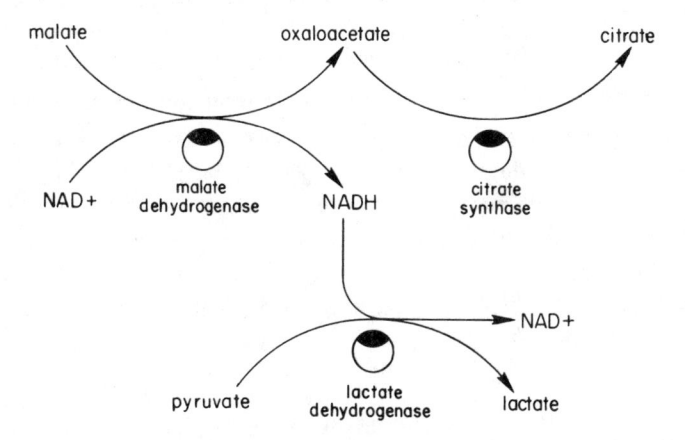

Figure 7.7. The co-immobilization of the three enzymes shown above results in a significant rate enhancement over the soluble enzyme system.

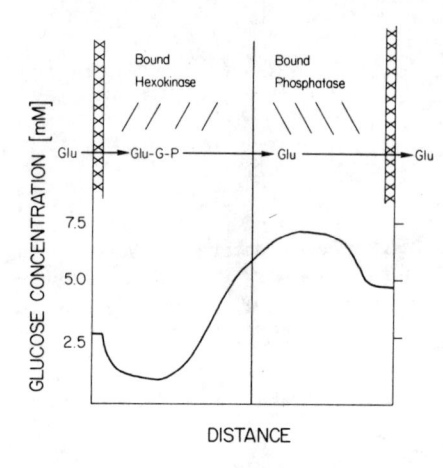

Figure 7.8. An artificial enzyme membrane that mimics active transport of glucose.

studies are relevant to the understanding of the control of the Krebs cycle. The operation of oxaloacetate as a regulating factor has been questioned because of the low level of oxaloacetate found in mitochondria (40 nm). Indeed, the concentration is so low that it is inconsistent with the known rate of the cycle (based on O_2 utilization) when parameters for soluble enzymes are used for the calculation. The results given above would explain this oxaloacetate level and support the idea of a compartmentalization of Krebs cycle enzymes into a cluster with a resulting kinetic advantage.

Broun and coworkers have described a bi-enzymic membrane consisting of hexokinase and a phosphatase that hydrolyzes glucose-6-phosphate (29). On the outside of the double enzyme membrane are selective membranes that restrict passage of glucose-6-phosphate (Fig. 7.8). In the presence of ATP, glucose can be transported against a concentration gradient: it enters on the left and is phosphorylated. The glucose-6-phosphate cannot leave the hexokinase layer to the left and tends to go to the right, where it is dephosphorylated to glucose by the phosphatase. The glucose may diffuse to the left, where it would be rephosphorylated, or it may diffuse out of the membrane by going to the right. The concentration profile would be sinusoidal as a result of interplay between diffusion and the enzymatic reactions. Notice that the glucose concentration on the right boundary is higher than that on the left.

D. APPLICATIONS

There are many existing and potential applications of bound enzymes. Large-scale industrial, synthetic, analytical and medical applications will be discussed here.

Large-Scale Industrial Application

Ichiro Chibata, at the Tanabe Company in Japan, is usually credited with the first industrial process based on the use of a bound enzyme (30). The process, shown in Figure 7.9, is for production of optically pure amino acids. The enzyme is L-amino acid acylase, which specifically removes an acetyl group from the L-isomer:

$$\text{D,L-AcNHCHRCOOH} \longrightarrow \text{L-NH}_2\text{CHRCOOH} + \text{D-AcNHCHRCOOH}$$

L-amino acid acylase

The acetyl D,L-amino acid is passed through the enzyme column and the free L-amino acid is crystallized. The remaining acetyl-D-amino acid is recycled after racemization (Fig. 7.9). The enzyme is immobilized by adsorption on DEAE-Sephadex. Chibata and Tosa have provided a report on the study of supports, immobilization of acylase, stability of enzyme conjugates, and economics of the process (30).

The process that involves the largest volume is the production of high fructose syrup through the use of immobilized glucose isomerase:

$$\text{cornstarch} \longrightarrow \text{glucose} \longrightarrow \text{fructose} + \text{glucose}$$
$$\uparrow$$
immobilized glucose isomerase

The fructose syrups are sweeteners that are widely used in soft drinks and other food products. It has been predicted that the corn-based products will command 50% of the U.S. sweetener market by the year 1990 (31).

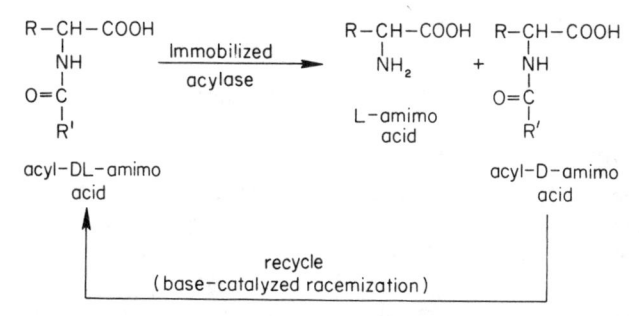

Figure 7.9. The production of optically pure amino acids, using immobilized acylase.

Immobilized β-galactosidase is used in the treatment of milk and milk products. β-galactosidase catalyzes the hydrolysis of lactose (milk sugar) to glucose and galactose. Many people are "lactose-intolerant," so a treatment to remove lactose from milk and milk products would be useful. Since whey contains large amounts of lactose, treatment with β-galactosidase facilitates the use of whey solids in food products.

Use of Bound Enzymes in the Synthesis of Fine Chemicals and Pharmaceuticals

Antibiotic Synthesis. Bound penicillin amidase is used industrially for the conversion of penicillin G to 6-amino penicillanic acid (32):

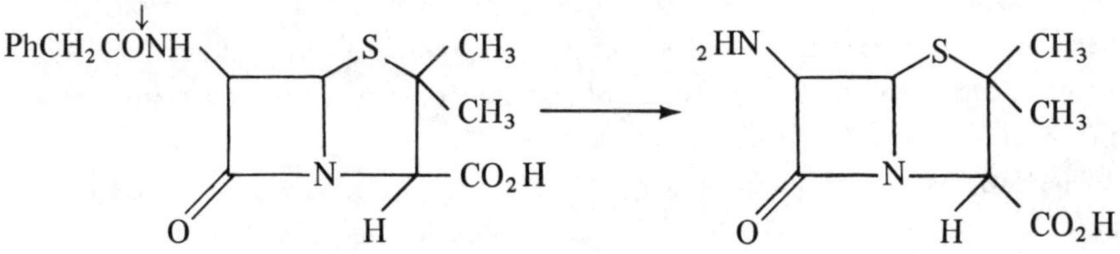

$$\text{Penicillin G} \longrightarrow \text{6-Amino Penicillanic Acid}$$

Konecny and Voser have worked on the use of an immobilized acetyl esterase in the synthesis of cephalosporins (33):

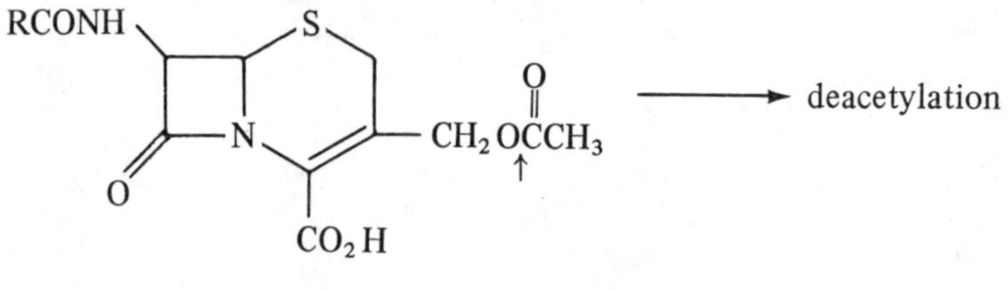

$$\longrightarrow \text{deacetylation}$$

Steroid Synthesis. An obvious area for the use of bound enzymes is in the stereospecific transformation of steroids. Since no byproducts are formed, economy is gained in the workup steps. An example is the conversion of compound S to prednisolone (34):

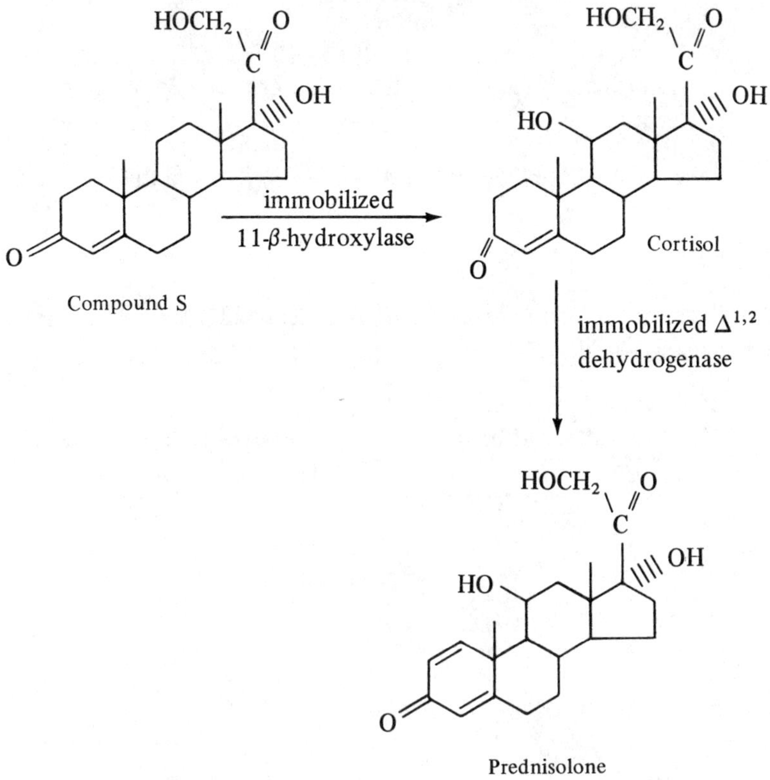

D. APPLICATIONS

Compound S →(immobilized 11-β-hydroxylase)→ Cortisol →(immobilized Δ^{1,2} dehydrogenase)→ Prednisolone

Peptide Synthesis. In my laboratory we have developed a method for sequential polypeptide synthesis that is based on the use of an immobilized enzyme. The peptide chain is grown by extension of the C-terminus, using carbodiimide in water with an excess of amino acid ethyl ester. The ethyl ester is hydrolyzed in the deblocking step by treatment with carboxypeptidase Y. This enzyme has negligible peptidase activity but maximal esterase activity at pH 8.5. Immobilization is accomplished by adsorption of the enzyme to immobilized concanavalin A, followed by a crosslinking step. In the synthesis, solubility of the growing peptide chain is provided by the use of carboxymethyl polyethylene-glycol (CM-PEG) as a handle. The synthetic procedure is summarized on the following page:

$$\text{CM-PEG-Gly-Met} \xrightarrow[\text{AA}_1\text{-OEt·HCl + EDC}]{\text{pH 6.0}} \text{CM-PEG-Gly-Met-AA}_1\text{-OEt}$$

$$\xrightarrow{\text{I-CPY (pH 8.5)}} \text{CM-PEG-Gly-Met-AA}_1\text{-OH} \xrightarrow{\text{iterate}} \text{CM-PEG-Gly-Met-}$$

$$\text{AA}_1 \ldots \text{AA}_n\text{-OEt} \xrightarrow{\text{CNBr}} \text{H-AA}_1 \ldots \text{AA}_n\text{-OEt}$$

Other Examples of Synthetic Applications. A list of additional synthetic applications appears in Table 7.3.

TABLE 7.3. Applications of Immobilized Enzymes in Organic Synthesis

Compound(s)	Immobilized Enzyme(s)
Antibiotics	
Penicillin	Penicillin amidase
Cephalosporin	Acetyl esterase
Steroids	
Cortisol	11-β-Hydroylase
Prednisolone	$\Delta^{1,2}$-Dehydrogenase
Peptide syntheses (deblocking)	Carboxypeptidase Y
(peptide bond formation)	Thermolysin, Trypsin, carboxypeptidase Y
Peptide ester synthesis	Chymotrypsin, subtilisin
Pyrrole porphobilinogen	δ-Aminolevulinate dehydratase
Chenodeoxycholate	3α- and 7-α-hydroxy steroid dehydrogenase
Keto acids	L-Amino acid oxidase
5'-Mononucleotides	5'-Phosphodiesterase, 5'-AMP deaminase

The Cofactor Problem. The problem of cofactor retention and regeneration is frequently encountered in synthetic uses of bound enzymes. Many enzymes use redox coenzymes (NAD, FAD) or cosubstrates (e.g., ATP), which must be regenerated and retained. In attempts to meet these requirements, a number of workers have studied polymer-bound NAD (e.g., 35). Regeneration may be accomplished by the use of alcohol dehydrogenase (ADH),

D. APPLICATIONS

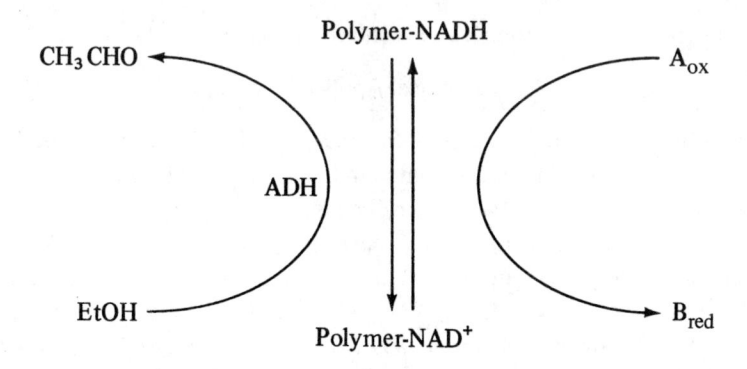

```
              Polymer-NADH
CH₃CHO  ←⎯⎯⎯⎯⎯↑⎯⎯⎯⎯⎯  A_ox
              │
          ( ADH )
              │
              ↓
EtOH    ⎯⎯⎯⎯⎯⎯⎯⎯⎯⎯→  B_red
              Polymer-NAD⁺
```

Larsson and collaborators have linked NAD directly to alcohol dehydrogenase through a covalent bond (36). Marikawa and coworkers have immobilized alcohol dehydrogenase, lactate dehydrogenase (LDH), and dextran-NAD together on the same support (37). The idea is to have the dextran-NAD serve two enzymes—one for the desired process, the other for regeneration. Lactate was produced without exogenous NAD. Electron flow probably occurs as follows:

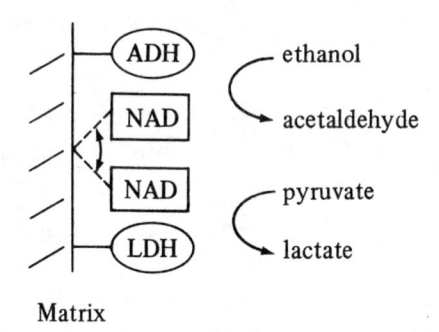

```
  │─( ADH )    ⌐ ethanol
  │            ⌐
  │  [ NAD ]   ⌐ acetaldehyde
  │     ⋮
  │  [ NAD ]   ⌐ pyruvate
  │            ⌐
  │─( LDH )    ⌐ lactate
  Matrix
```

A cofactor required in many synthetic applications is adenosine triphosphate (ATP). An enzymatic method for regeneration of ATP from AMP has been developed by Whitesides' group (38). The scheme employs acetyl phosphate (AcP) as the phosphate donor and two immobilized kinases—adenylate kinase and acetate kinase:

$$\text{AMP} + \text{ATP} \xrightarrow{\text{adenylate kinase}} 2\,\text{ADP}$$

$$2\,\text{ADP} + 2\,\text{AcP} \xrightarrow{\text{acetate kinase}} 2\,\text{ATP} + 2\,\text{acetate}$$

Using this system, bench-top reactors can produce at least 1 g of ATP/h.

Analytical Applications

Enzymes are used widely as analytical tools because of their exact specificity. The amount of a given chemical compound in a complex physiological fluid—or in plant and municipal waste streams—can often be determined enzymatically with little or no pretreatment of the sample. The possibility of enzyme reuse and continuous, automatable monitoring devices makes immobilization attractive. Updike and Hicks introduced an "enzyme electrode" for the determination of glucose (39). An oxygen electrode was coated with a layer of polyacrylamide that contained the enzyme glucose oxidase. The glucose in the test sample is oxidized to gluconic acid:

$$\text{glucose} + O_2 \xrightarrow{\text{glucose oxidase}} \text{gluconic acid} + H_2O_2$$

A similar enzyme electrode was developed for measuring urea concentration (40). In this case an ammonia-sensitive electrode and the enzyme urease were employed:

$$\text{urea} + H_2O \xrightarrow{\text{urease}} 2\,NH_3 + CO_2$$

Many analytical methods based on bound enzymes have followed the first enzyme electrodes (41). Thermal detectors have created much interest because of their sensitivity and applicability to many different immobilized enzyme systems (42). Weaver and collaborators have been attempting to combine the specificity of enzymes with the great sensitivity of the mass spectrometer (43). In this approach the enzyme catalyzes a reaction that produces a volatile product, which enters the mass spectrometer through a semipermeable membrane.

Medical Applications

The possible use of immobilized enzymes as therapeutic agents has been the subject of study in many laboratories. But the use of soluble enzymes in treatment of disease has a very serious drawback: the injection of foreign proteins into the blood can result in life-threatening allergic reactions and/or rapid deactivation of the enzyme. The concept of a bound enzyme is appealing because the foreign protein would presumably be less susceptible to attack by the immume systems of the body or by proteases in the blood. Treatment of kidney disease with bound urease and of the asparagine-dependent lymphosarcomas with bound asparaginase has been investigated. A serious problem in this work centers on development of biocompatible supports that deliver high amounts of enzyme for long periods. An extensive treatise on this subject has recently appeared (44).

REFERENCES

1. T. Tosa, T. Mori, N. Fuse, and I. Chibata, *Agric. Biol. Chem.* **33**, 1047 (1969).
2. B. H. J. Hofstee, *Biochem. Biophys. Res. Commun.* **53**, 1137 (1973).
3. L. G. Butler, *Arch. Biochem. Biophys.* **171**, 645 (1975).
4. R. Chambers, W. Cohen, and W. H. Baricos, *Methods Enzymol.* **44**, 291 (1976).
5. W. Marconi, S. Gulinelli, and F. Morisi, in *Insolubilized Enzymes* (M. Salmona, C. Saronia, and S. Garatini, Eds.), Raven Press, New York (1974), p. 51.
6. T. M. S. Chang, in *Biomedical Applications of Immobilized Enzymes and Proteins,* Vol. 1 (T. M. S. Chang, Ed.), Plenum, New York (1977), p. 69.
7. P. Bernfeld and J. Wan, *Science* **142**, 678 (1963).
8. A. C. Koch-Schmidt, in *Biomedical Applications of Immobilized Enzymes and Proteins,* Vol. 1 (T. M. S. Chang, Ed.), Plenum, New York (1977), p. 48.
9. L. Goldstein and G. Manecke, in *Applied Biochemistry and Bioengineering* (L. B. Wingard, E. Katchalski-Katzir, and L. Goldstein, Eds.) Academic Press, New York (1976), p. 23.
10. J. Kahn and W. Wilchek, *Biochem. Biophys. Res. Commun.* **84**, 7 (1978).
11. H. H. Weetall, *Science* **166**, 615 (1969).
12. J. Konecny and W. Voser, *Biochem. Biophys. Acta.* **484**, 367 (1972).
13. F. A. Quiocho, in *Insolubilized Enzymes* (M. Salmona, C. Saronio, and S. Garatinni, Eds.) Raven, New York (1974), p. 113.
14. G. P. Royer and R. Uy, *J. Biol. Chem.* **248**, 2627 (1973).
15. D. Gabel, I. Z. Steinberg, and E. Katchalski, *Biochemistry* **10**, 4661 (1971).
16. L. J. Berliner, S. T. Miller, R. Uy, and G. P. Royer, *Biochem. Biophys. Acta.* **315**, 195 (1973).
17. G. P. Royer, W. E. Swartz, and F. A. Liberatore, *Methods Enzymol.* **47**, 40 (1977).
18. L. Goldstein, Y. Levin, and E. Katchalski, *Biochemistry* **3**, 1913 (1964).
19. C. M. Wharton, E. M. Crook, and K. Brocklehurst, *Eur. J. Biochem.* **6**, 572 (1968).
20. L. Goldstein, *Methods Enzymol.* **44**, 402 (1976).
21. H. L. Brockman, J. H. Law, and F. J. Kedzy, *J. Biol. Chem.* **248**, 4965 (1973).
22. J. M. Engasser and C. Horvath, in *Applied Biochemistry and Bioengineering* (L. B. Wingard, E. Katchalski-Katzir, and L. Goldstein, Eds.), Academic Press, New York (1976), p. 142.
23. W. E. Hornby. M. D. Lilly, and E. M. Crook, *Biochem. J.* **98**, 420 (1966).

24. R. Goldman, D. Kedem, and E. Katchalski, *Biochemistry* **7**, 4518 (1968).
25. P. S. Bunting and K. J. Laidler, *Biochemistry* **11**, 4477 (1972).
26. K. Mosbach and B. Mattiasson, *Acta Chem. Scand.* **24**, 2093 (1970).
27. B. Mattiasson and K. Mosbach, *Biochem. Biophys. Acta* **235**, 253 (1971).
28. P. A. Srere, B. Mattiason, and K. Mosbach, *Proc. Natl. Acad. Sci. USA* **70**, 2534 (1973).
29. G. Broun, D. Thomas, and E. Selegny, *J. Membrane Biol.* **8**, 313 (1972).
30. I. Chibata and T. Tosa, in, *Applied Biochemistry and Bioengineering* (L. B. Wingard, E. Katchalski-Katzir, and L. Goldstein, Eds.), Academic Press, New York (1976), p. 329.
31. C. E. Kean, *Food Product Devel.*, April, 43 (1978).
32. W. Marconi, F. Cecere, F. Morisi, G. D. Penna, and B. Rapperole, *J. Antibiotics* **26**, 228 (1973).
33. J. Konecny and W. Voser, *Biochem. Biophys. Acta* **484**, 367 (1972).
34. K. Mosbach and P. O. Larsson, *Biotech. Bioeng.* **12**, 19 (1970).
35. P. O. Larsson and K. Mosbach, *FEBS Lett.* **46**, 119 (1974).
36. P. O. Larsson, K. Mosbach, and J. D. Mansson, *Eur. J. Biochem.* **86**, 455 (1978).
37. Y. Marikawa, I. Karube, and S. Suzuki, *Biochem. Biophys. Acta* **523**, 263 (1978).
38. C. R. Gardner, C. K. Colton, R. S. Lauger, B. K. Hamilton, M. C. Archer, and G. M. Whitesides, *J. Am. Chem. Soc.* **101**, 5829 (1979).
39. S. J. Updike and G. P. Hicks, *Nature* (London) **214**, 986 (1967).
40. G. G. Guilbault and J. G. Montalvo, *J. Am. Chem. Soc.* **92**, 2533 (1970).
41. L. D. Bowers and P. W. Carr, *Analyt. Chem.* **48**, 544A (1976).
42. B. Danielson and K. Mosbach, *Methods Enzymol.* **44**, 667 (1976).
43. J. C. Weaver, C. L. Cooney, S. R. Tannenbaum, and S. P. Fulton, in *Biomedical Applications of Immobilized Enzymes and Proteins*, Vol. 2 (T. M. S. Chang, Ed.) Plenum, New York (1977), p. 191.
44. *Biomedical Applications of Immobilized Enzymes and Proteins*, Vols. 1 and 2 (T. M. S. Chang, Ed.), Plenum, New York (1977).

CHAPTER EIGHT

ENZYME-LIKE SYNTHETIC CATALYSTS "SYNZYMES"

A standard approach in the study of biologically active compounds includes three steps: purification, structure determination, and synthesis from nonbiological materials. Demonstration of the biological activity of a vitamin or hormone of synthetic origin has served as the ultimate proof of structure for many compounds. Would this approach work for a molecule as large and complex as an enzyme? The basic knowledge gained from such studies and the obvious practical benefit of a successful attempt have led a number of people to try to mimic enzymes with organic molecules of varying degrees of complexity.

What would constitute a success? Enzymes, as stated earlier, possess three remarkable properties: spectacular catalytic ability, specificity, and sensitivity to control. The first trait to mimic would be catalytic efficacy. Since natural enzymes produce rate enhancements of 10^6 to 10^{14}, a near miss in the effort to prepare a synzyme might yield a fabulous commercial success if the synthetic catalyst were *inexpensive* and *stable*.

Enzymatic rate enhancement stems from multifunctional catalysis, binding of substrate, microenvironmental (medium) effects, and strain. In this chapter several types of model systems that exhibit these traits are discussed.

A. MACROCYCLES

Cycloamyloses are naturally occurring cyclic oligosaccharides composed of D-glucopyranose rings in α-1,4 linkage (Fig. 8.1). The three most thoroughly studied cycloamyloses are those with six, seven, and eight glucose units (1). These structures are called cyclohexaamylose (α-cyclodextrin), cycloheptaamylose (β-cyclodextrin), and cyclooctaamylose (γ-cyclodextrin), respectively. The cycloamyloses are toroidal in shape, with an apolar interior cavity. The lining of the cavity is comprised of CH groups and glycosidic oxygens; its diameter ranges from 4.5 Å. to 8.5 Å. The primary hydroxyls (of C-6 on the glucopyranose ring) are on one end and the secondary hydroxyls (of C-2 and C-3 on the glucopyranose) are on the other. The interest in these molecules as enzyme models derives from the fact that inclusion complexes are formed between cyclodextrins and apolar molecules (2, 3). Such complexes, with *t*-butylphenyl acetates, appear in Figure 8.2. Acyl transfer from substituted phenyl esters to the secondary hydroxyl of the cyclodextrins is fast, compared to the rate of acyl transfer to water. Preference for meta-substituted nitrophenyl esters and a binding step were demonstrated (3). The deacylation reaction is slow. When the substrate is bound by the acyl group preferentially, an enormous rate of acyl transfer is observed (>50,000 times the background hydrolysis rate; 4). Czarnieki and Breslow demonstrated this with the *p*-nitrophenyl ester of ferrocinnamic acid:

The K_D (analogous to K_S) was found to be 7 mM for this system. Turnover of the acylcycloamylose was not reported.

Cycloamylose-(N,N'-dimethylaminoethyl) acetohydroxamic acid (1) has proved to be a true catalyst in that it deacylates rapidly at neutral pH (1).

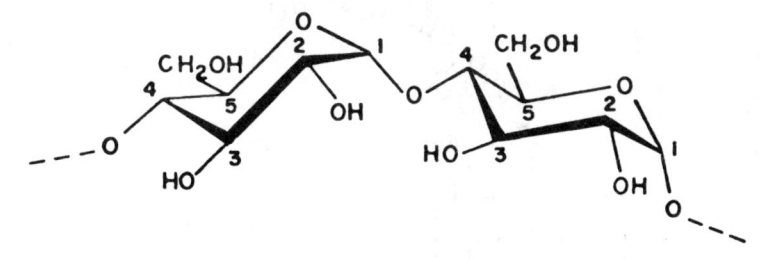

Figure 8.1. The α-1, 4 linkage of the cyclodextrins.

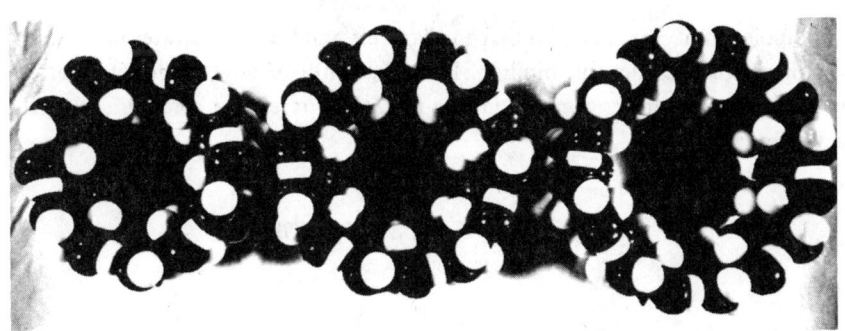

Figure 8.2. (Top) The inclusion complex of cyclohexamylose with *m-t*-butylphenyl-acetate. (Bottom) Space-filling models of cyclohexaamylose, cycloheptaamylose, and cyclooctaamylose (left to right).

$$\text{OCH}_2\text{CONCH}_2\text{CH}_2\text{N}(\text{CH}_3)_2$$
$$\overset{|}{\text{OH}}$$

|||||| cavity ||||||

1

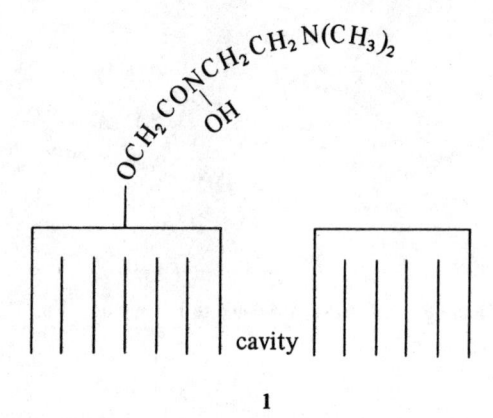

In this case the nucleophile for esterolysis is the acetohydroxamic acid group, rather than the secondary hydroxyl group of the cyclodextrin. The tertiary amine acts as a general base.

Discovery of the catalytic properties of cyclodextrins has stimulated the preparation of synthetic macrocycles that form inclusion complexes. A cyclic amine with hydroxamate groups, 2, was prepared by Hershfield and Bender (5).

$$\underset{\underset{\text{CH}_3}{|}}{\overset{\overset{\text{O}}{\|}}{\text{HONCCH}_2}}\text{N} \genfrac{}{}{0pt}{}{\diagup (\text{CH}_2)_{12} \diagdown}{\diagdown (\text{CH}_2)_{12} \diagup} \text{N}\text{CH}_2\underset{\underset{\text{CH}_3}{|}}{\overset{\overset{\text{O}}{\|}}{\text{C}}}\text{NOH}$$

2

The apolar cavity of **2** is 5 to 6 Å in diameter and binds substrates in a manner analogous to cyclodextrin binding. For apolar nitrophenyl carboxylate esters, large rate accelerations were produced by **2**. The reference compound was the acyclic analog, **3**.

$$[(\text{CH}_3)_2\text{CHCH}_2]_2\text{NCH}_2\underset{\underset{\text{CH}_3}{|}}{\overset{\overset{\text{O}}{\|}}{\text{C}}}\text{NOH}$$

3

For the hydrolysis of *p*-nitrophenyl laurate a rate enhancement of 7600 was found.

Paracyclophanes have been made with pendant catalytic groups (6). The cavity of compound **4** is about 6.5 Å. The hydrolysis rates of *p*-nitrophenyl esters of long-chain fatty acids are accelerated by **4**, but the hydrolysis rates of esters of short-chain fatty acids are not changed. A more elaborate paracyclophane, **5**,

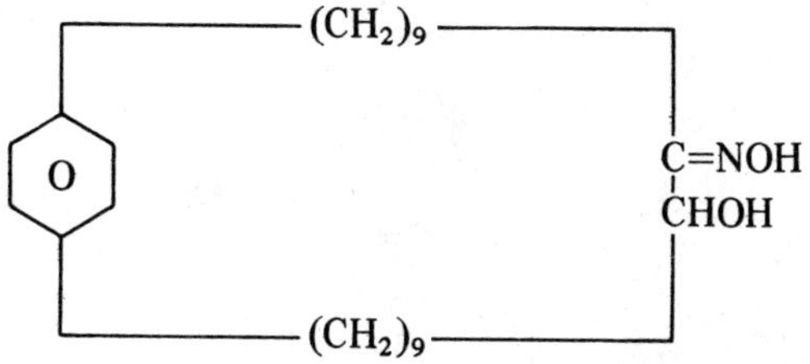

4

has been reported (7). Two functional groups, a binding site and a metal ion, make up the "active center" of this catalyst. Although rate enhancements were small for this system, evidence for the metal ion's acting to polarize the carbonyl oxygen of the bound ester was reported. Electrophilic catalysis by the metal ion permits the protonated oxime to perform as a nucleophile.

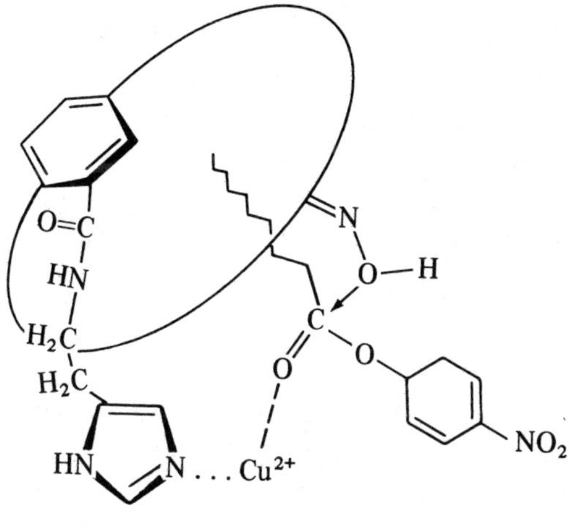

5

B. CATALYSTS BASED ON SYNTHETIC POLYMERS

The effects of polymers on reactions in solution have been studied for several decades. Nonspecific electrostatic effects were investigated by Morawetz (8). Polyions inhibit reactions between ions of opposite charge. Consider the reaction of a positively charged ester with hydroxide ion in the presence of a negatively charged polymer (reaction 1).

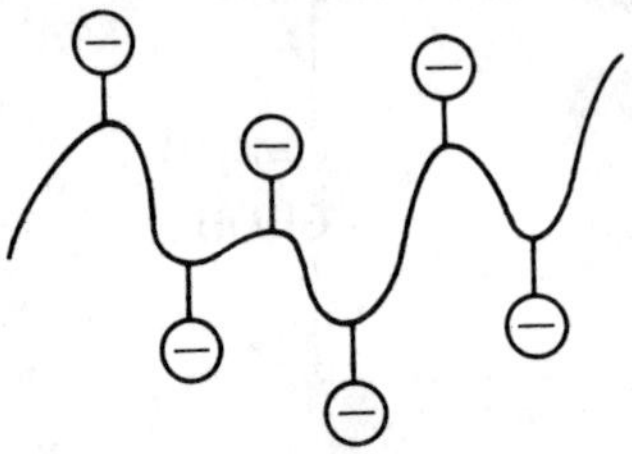

$$\text{polyanion} + R^+CO_2R + OH^- \longrightarrow \quad (1)$$

The polymer will bind the ester, but the OH⁻ species would be repelled, which results in a retardation of reaction rate. In contrast, reactions between ions of like charge would be accelerated by a polymer of opposite charge. Sodium poly-(vinylsulfonate) accelerates reaction 2 by a factor of 176,000.

$$Co(NH_3)_5Cl^+ + Hg^{2+} + H_2O \rightarrow Co(NH_3)_5H_2O^{3+} \quad (2)$$
$$\uparrow \qquad \qquad + HgCl^+$$
$$(-CH_2\underset{|}{CH}-)_n$$
$$SO_3^-Na^+$$

In this case the reactants are concentrated in the vicinity of the polymer, which results in an impressive increase in the reaction rate.

Letsinger and Savereide studied catalysis by a vinyl polymer with nucleophilic groups (9). They demonstrated that poly-(4-vinylpyridine) was an active catalyst in the solvolysis of 3-nitro-4-acetoxy-benzenesulfonate, **6**, in 50% ethanol.

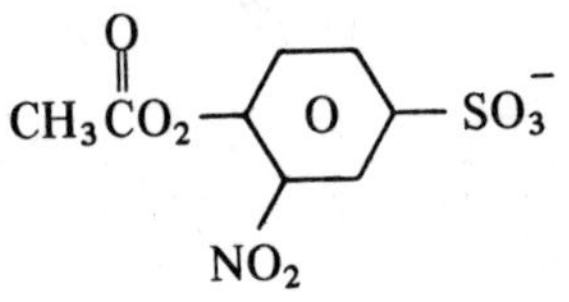

$$CH_3CO_2-\underset{NO_2}{\underset{|}{\bigcirc}}-SO_3^-$$

6

The rate enhancement results from electrostatic binding of the negatively charged substrate to the polymer, which is partially protonated. The greatest rate occurs when 60% of the pyridine residues are deprotonated ($\alpha = 0.6$, Fig. 8.3).

Overberger and Salamone studied various polymers resulting from polymerization of 4(5)-vinyl imidazole (10). With the homopolymer of 4(5)-vinyl imidazole, three types of cooperative effect were noted: imidazole acting as a nucleophile with a nearby imidazole anion acting as a general base, imidazole acting as a nucleophile with a neighboring neutral imidazole acting as a base, and attraction of the negatively charged substrate by an imidazole cation near a neutral imidazole, which acts as a nucleophile in analogy to catalysis by poly-(4-vinyl pyridine). A copolymer of 4(5)-vinyl imidazole and 4-vinyl phenol produced significant rate enhancements in the hydrolysis of nitrophenyl esters. At the pH value at which 10% of the phenol residues were ionized, the polymer reacted with p-nitrophenyl acetate with a rate constant of 150 M^{-1} min^{-1}, compared to 2.3 M^{-1} for imidazole. The copolymer of vinyl imidazole and methoxystyrene exhibited no special

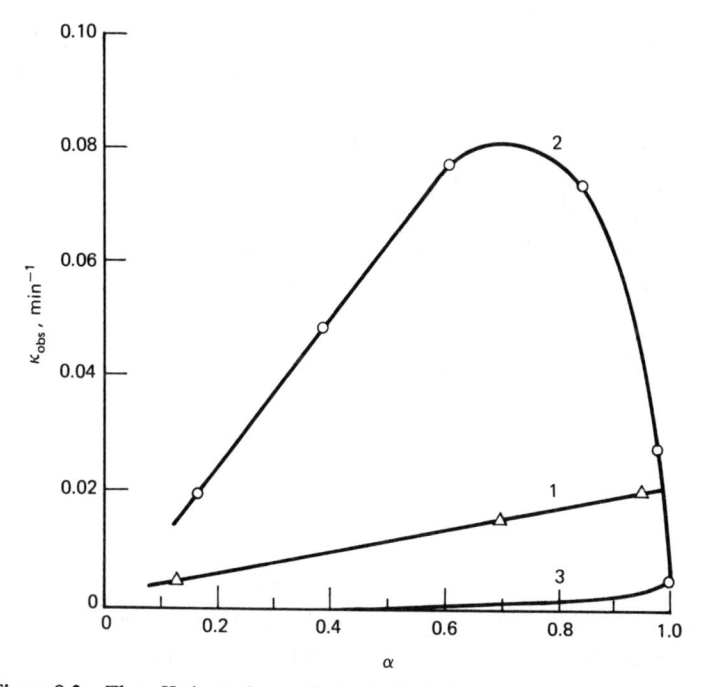

Figure 8.3. The pH dependence of the hydrolysis of 3-nitro-4-acetoxy-benzenesulfonate, as catalyzed by poly-(4-vinlypyridine).

properties. The proposed mechanism of action of the phenol-imidazole polymer involved the phenolate ion, as follows:

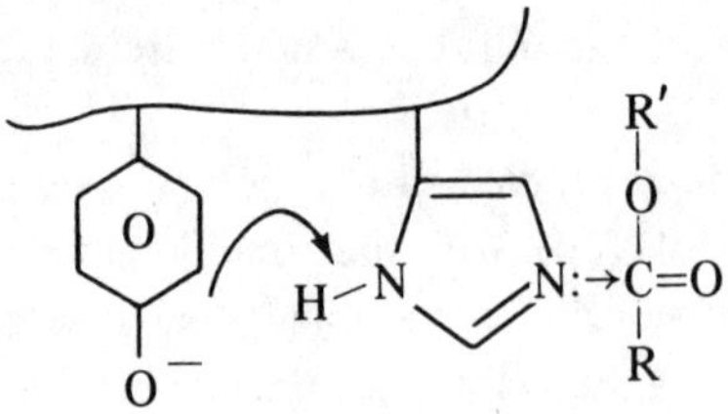

$$\text{[phenolate-imidazole-ester mechanism diagram]}$$

The hydroxamate group, although not a constituent of natural enzymes, has been of interest because of its high nucleophilicity. Compounds 7 and 8 have been studied by Gruhn and Bender (11, 12) and by Kunitake and coworkers (13).

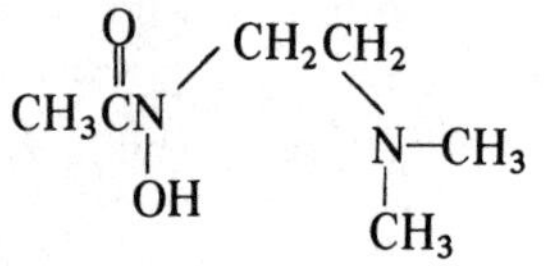

$$\text{7}\qquad\text{8}$$

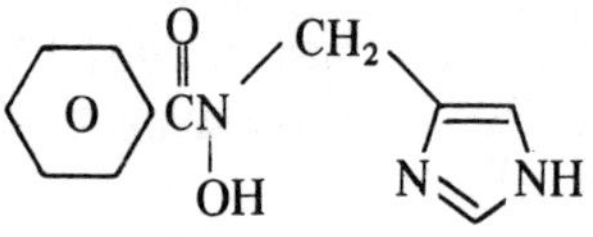

Deacylation of **8** proceeds 13 times faster than the analog without imidazole. In addition, a solvent deuterium isotope effect of 2 was observed, which suggests that the imidazole groups assist deacylation by acting as a general base. For a polymeric catalyst with the segment **9**, the deacylation rate is 100 times greater than that for the corresponding reaction of the monomeric analog (14).

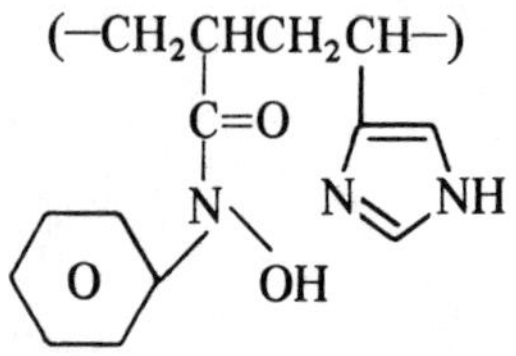

$$\text{9}$$

Since linear polymers are generally not capable of binding small molecules with affinities characteristic of enzymes, Klotz and coworkers have concentrated on a globular synthetic polymer, poly(ethyleneimine) and, to a lesser extent, on crosslinked polylysine. Poly(ethyleneimine) (PEI) is a highly branched, water-soluble polymer that results from the acid-catalyzed polymerization of ethyleneimine (aziridine) (15). The chemical structure can be represented as

$$_2HN-(-CH_2-CH_2-NH)_x-(-CH_2-CH_2-N)_y-CH_2-CH_2-NH_2$$
$$|$$
$$CH_2$$
$$|$$
$$CH_2$$
$$|$$
$$N$$
$$|$$

The branching may be depicted as

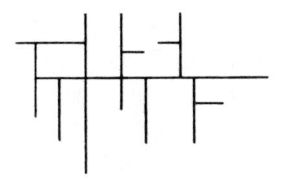

The amine distribution is 25% primary, 50% secondary, and 25% tertiary.

Klotz and collaborators showed that PEI with pendant apolar groups possessed tremendous ability to bind apolar molecules of low molecular weight (16; Fig. 8.4). Royer and Klotz demonstrated that lauroyl PEI reacted very

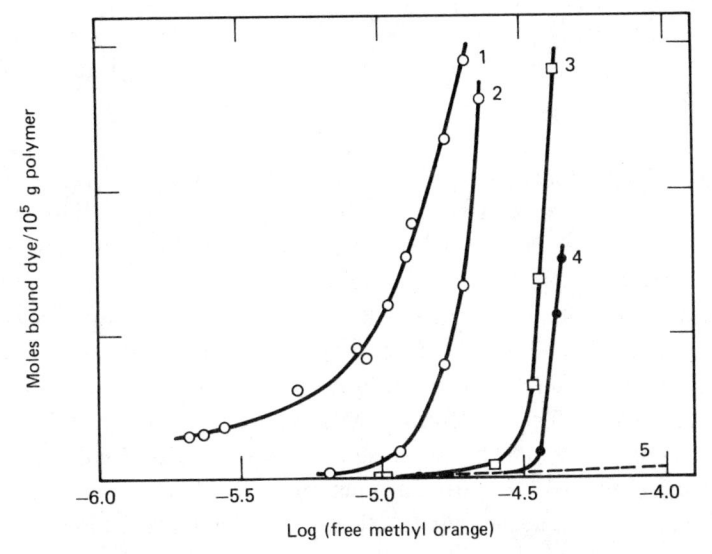

Figure 8.4. The binding of methyl orange by apolar derivatives of poly (ethyleneimine).

rapidly with nitrophenyl esters (Table 8.1; 17). With *p*-nitrophenyl laurate as substrate, the enhancement of polymer rate over the propylamine rate is 10^4. A substantial part of this acceleration presumably results from complexation of the apolar substrate by the apolar groups on the polymer. This reaction is really a one-step aminolysis with no turnover.

The next step in the elaboration of a PEI-based catalyst was the introduction of catalytic groups, along with the binding groups. The alkylation of PEI with chloromethyl imidazole and dodecyl iodide produced **10** which is a true catalyst for hydrolysis of carboxylate and sulfate esters.

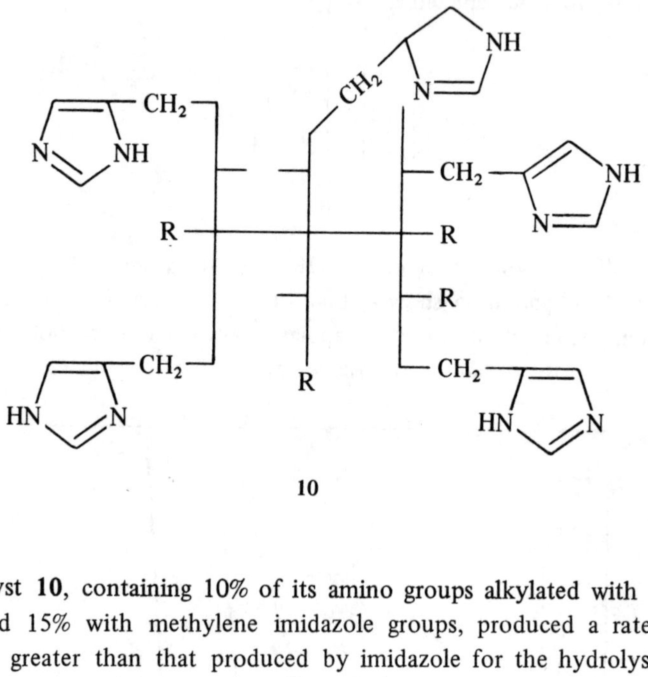

10

The catalyst **10**, containing 10% of its amino groups alkylated with dodecyl groups and 15% with methylene imidazole groups, produced a rate nearly 300 times greater than that produced by imidazole for the hydrolysis of *p*-nitrophenyl caproate at pH 7.3 and 25°C (18).

Kiefer and his coworkers examined **10** as a catalyst for the hydrolysis of 4-nitrocatechol sulfate (19). Unlike the corresponding carboxylate ester, the sulfate ester is stable at room temperature. In fact, 2 M imidazole produces no detectable hydrolysis at 20°C. By contrast, **10** is a very effective catalyst. The rate enhancement produced by **10** is even greater than that produced by the enzyme aryl sulfatase. Compound **10** is 10^{12} times better than imidazole as a catalyst for the hydrolysis of 4-nitrocatechol sulfate. Michaelis-Menten kinetics and a double-

TABLE 8.1. First-Order Rate Constants for Amine Acylation by p-Nitrophenyl Esters[a] (17)

Amine	p-Nitrophenyl Acetate	k X 10² min[b] p-Nitrophenyl Caproate	p-Nitrophenyl Laurate
Propyl	0.98	0.51	0.053
PEI-6[c]	3.60	1.47	0.11
PEI-18[c]	4.38	1.57	0.11
PEI-600[c]	4.60	1.80	0.17
L (10%) – PEI-6[d]	15.2	68.1	698

[a]Measurements made at pH 9.0 in 0.02 M tris (hydroxymethyl)-aminomethane buffer, 25°. Stock solutions of substrate were made in acetonitrile; hence the final buffer also contained 6.7% acetonitrile.
[b]$k = k_a$, where k_a is the measured rate constant in the presence of amine and k_0 is that for the hydrolysis in tris buffer alone; k_0 is 0.94 × 10⁻² min⁻¹ for the acetyl ester, 0.61 × 10⁻² min⁻¹ for the caproyl ester, and 0.023 × 10⁻² min⁻¹ for the lauroyl ester.
[c]The numeral following "PEI" multiplied by 100 is the molecular weight of the polymer sample.
[d]This sample of PEI-6 has 10% of its nitrogens acylated with lauroyl groups.

displacement pathway were observed (19). In addition to a strong and productive catalyst-substrate interaction, a medium effect may be a significant factor here. For the hydrolysis of alkyl hydrogen sulfates, Batts reported a large solvent effect (20): moist dioxane was better than water by a factor of 10^7! It is not unreasonable to conclude that 10 provides a microenvironment that resembles moist dioxane.

Reactions other than ester hydrolysis have been studied with PEI derivatives as catalysts. Dodecyl-PEI with free amino groups and quaternized nitrogens, dodecyl-PEI-Q-NH$_2$, effectively catalyzes the decarboxylation of oxaloacetic acid:

$$HO_2CCCH_2CO_2^- + \text{dodecyl--PEI--Q--NH}_2 \rightarrow CH_3CCO_2H$$
$$\underset{O}{\|} \qquad\qquad\qquad\qquad\qquad\qquad\qquad \underset{O}{\|}$$

$$+ CO_2 + \text{dodecyl--PEI--Q--NH}_2$$

At pH 4.5 the polymer is 10^5 times more effective than ethylamine (21). Dodecyl-PEI with fully quaternized nitrogens (no free amines) was found to be an effective catalyst for the decarboxylation of nitrobenzisoxazolecarboxylic acid,

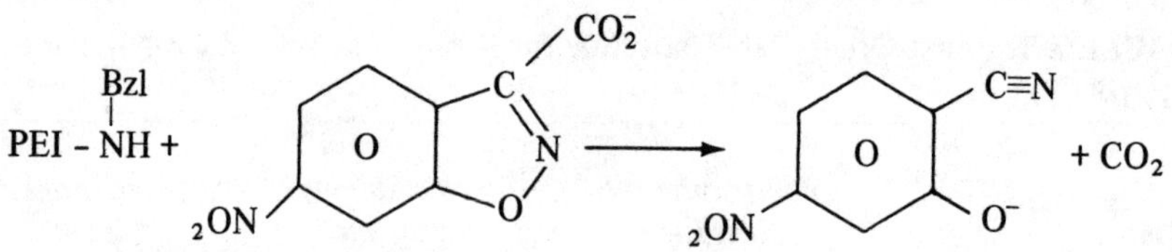

[Reaction scheme showing Bzl-PEI-NH+ attacking a benzisoxazole with CO2- and 2ON substituents, yielding a product with C≡N group and + CO2]

Strong binding and stabilization of delocalized charge on the transition state were cited to account for a rate enhancement of 10^3 over background (22). Weatherhead and collaborators have shown that PEI, benzylated to the extent of 10% of the amine residues, effectively cleaves Ellman's reagent [5,5'–dithiobis (2-nitrobenzoic acid)]:

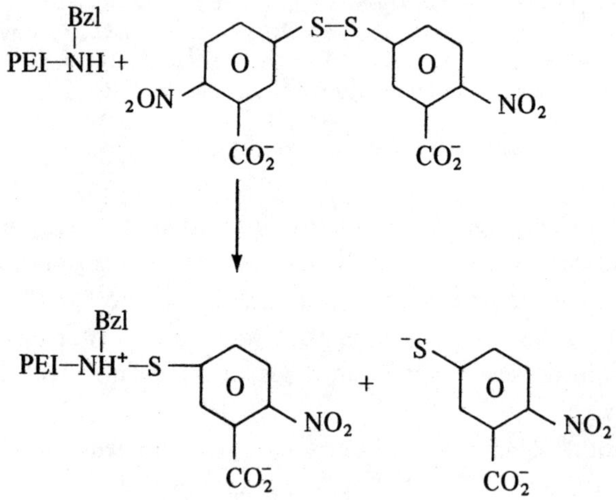

[Reaction scheme showing Bzl-PEI-NH+ attacking Ellman's reagent (disulfide with 2ON, NO2, CO2- substituents), yielding PEI-NH+-S-aryl and -S-aryl products]

A large rate enhancement (10^6) was attained, using an ideal simple amine as a reference compound. The apparent pK_a of Bzl-PEI was found to be 7.49. The rate constant for the reference amine with the same pK_a was taken from a Brønsted plot (24).

C. IMMOBILIZED ENZYME-LIKE CATALYSTS

There are three basic reasons for attempting to prepare insolubilized catalysts with enzyme-like properties. First, synthetic procedures would be simplified if the catalyst were built stepwise on a solid support. Second, reactive groups fixed on the matrix, such as thiols or metal ions, could not interact with one another

C. IMMOBILIZED ENZYME-LIKE CATALYSTS

to form binuclear structures, which are inactive. Third, if a commercially important enzyme-like catalyst is ever made, it will almost certainly be used in a heterogeneous system.

Breslow and his coworkers have prepared an insoluble cyclodextrin matrix by crosslinking cyclohexamymlose with epichlorohydrin (25). The resin was used in a packed bed for chlorination of anisole as follows: the cyclodextrin cavities were filled with anisole; aqueous HOCl was passed through the column; and the product, 99% p-chloroanisole, was eluted with tetrahydrofuran.

Meyers and Royer prepared the solid-phase analog of 10 (26). In the course of this work, a new support material was developed. Polymer "ghosts" are made in a three-step process (Fig. 8.5). In the first step the polymer is adsorbed onto beads of a porous, inorganic material, such as alumina (Fig. 8.5a). Crosslinking is the second step (Fig. 8.5b). In the final step (Fig. 8.5c), the alumina core is dissolved away to yield a hollow "ghost." The density of the wall structure and thickness of the wall can be controlled by the geometry of the inorganic beads, the amount of polymer adsorbed, and the degree of crosslinking. PEI "ghosts,"

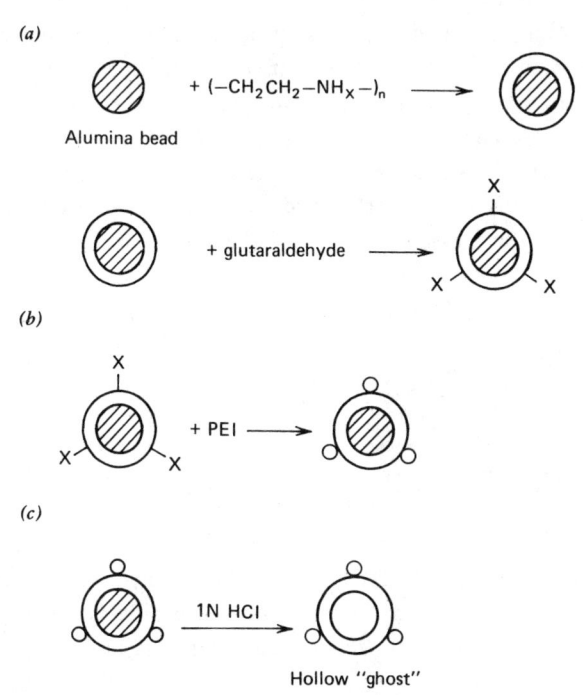

Figure 8.5. The preparation of polymer "ghosts."

crosslinked with glutaraldehyde, are rigid structures that are compatible with many solvents, including water. They are mechanically and chemically stable. PEI "ghosts" were modified with histidyl groups and lauroyl groups. The resulting analog of **10** was tested against p-nitrophenyl caproate as substrate. After a rapid initial phase, the reaction rate levels off at a point a third of the way to completion. In the final slow stage, the reaction rate in the presence of polymer "ghosts" actually falls below the background rate. Strong binding of substrate in areas inaccessible to water and/or product inhibition could explain this observation. The hydrolysis of p-nitrophenyl trifluoroacetanilide (**11**) proceeded with more conventional kinetics. In this case the leaving group is neutral rather than negatively charged.

$$CF_3CONH-\phi-NO_2$$

11

At pH 8.2 the lauroyl, -histidyl "ghosts" produced a rate 230 times greater than the imidazole-catalyzed rate. Constant activity after repeated treatments with excess substrate showed that the catalyst was regenerated.

A palladium catalyst with superior activity has been prepared with PEI "ghosts" as a starting material (27). Pd^{2+} was coordinated to the amines of PEI and then reduced with sodium borohydride. The Pd-PEI catalyst is very effective in the hydrogenolytic cleavage of the carbobenzoxy (Cbz) group with formic acid as the hydrogen donor:

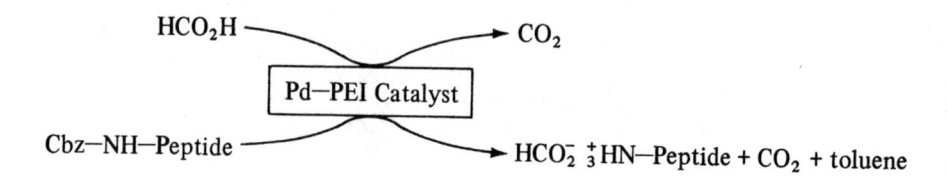

$HCO_2H \longrightarrow CO_2$
Pd—PEI Catalyst
$Cbz-NH-Peptide \longrightarrow HCO_2^- \; {}^+_3HN-Peptide + CO_2 + toluene$

The rate enhancement shown in Figure 8.6 may result from the attraction of reactants and/or repulsion of the product which is positively charged. However, enzyme-like behavior has not yet been proved in this system.

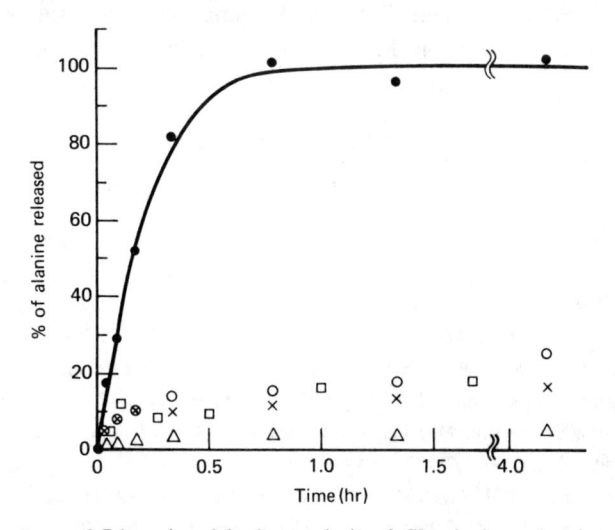

Figure 8.6. Rate of Pd-catalyzed hydrogenolysis of Cbz-alanine using formic acid as the hydrogen donor. Pd–PEI "ghosts," 100 mg (●); Aldrich 10% Pd on charcoal, lot no. 031597, 10 mg (□); Research inorganic/organic chemical 10% Pd on charcoal, two different samples, 10 mg (X and O); Pd–black, 1 mg (△).

D. CONCLUSIONS AND PROSPECTS

Many of the basic elements of enzymatic catalysis have been demonstrated in synthetic catalysts during the last 20 years. Substrate complexation and saturation have been shown frequently in diverse synthetic models. Nucleophilic catalysis and covalent intermediates have been demonstrated. Examples of cooperation among two or more catalytic groups are common. Medium effects have been amply demonstrated. Many of the "catalytic" systems discussed here, however, are still of no industrial interest, because only one or perhaps two properties of enzymes are duplicated. Some of the cyclodextrin systems, for example, show very rapid acylation but very slow deacylation. Unfortunately, the hydrolase models do not work on "hard" esters or amides. Specificity in the systems involving polymers is rather broad. However, this problem is not serious from a practical viewpoint. An enzyme-like synthetic catalyst need not function in a dynamic and complex biological milieu.

In nature, a billion years were required for the evolution of enzymes. In about 25 years, many enzymatic properties have been duplicated in human-made catalysts. Although a true "synzyme" of practical importance has not appeared, it would not surprise me to see one in operation before the year 2000.

REFERENCES

1. M. L. Bender and M. Komiyama, *Cyclodextrin Chemistry*, Springer-Verlag, New York (1977).
2. F. Cramer, W. Saenger, and H. C. Spatz, *J. Am. Chem Soc.* 89, 14 (1967).
3. R. L. Van Etten, J. F. Sebastian, G. A. Clowes, and M. L. Bender, *J. Am. Chem. Soc.* 89, 3242 (1967).
4. M. F. Czarnieki and R. Breslow, *J. Am. Chem. Soc.* 100, 7771 (1978).
5. R. Herschfield and M. L. Bender, *J. Am. Chem. Soc.* 94, 1376 (1972).
6. Y. Murakami, J. Sunamoto, and K. Kano, *Bul. Chem. Soc. Jap.* 47, 1238 (1974).
7. Y. Murakami, Y. Aoyama, M. Kida, and J. Kirkuchi, *J.C.S. Chem. Comm.* 494 (1978).
8. H. Morawetz, *Acc. Chem. Res.* 3, 354 (1970).
9. R. L. Letsinger and T. J. Savereide, *J. Am. Chem. Soc.* 84, 3122 (1962).
10. C. G. Overberger and J. C. Salamone, *Acc. Chem. Res.* 2, 217 (1969).
11. W. B. Gruhn and M. L. Bender, *J. Am. Chem. Soc.* 91, 5883 (1969).
12. W. B. Gruhn and M. L. Bender, *Bioorg. Chem.* 4, 219 (1975).
13. T. Kunitake, Y. Okahata, and T. Tahara, *Bioorg. Chem.* 5, 155 (1976).
14. T. Kunitake and Y. Okahata, *Macromolecules* 9, 15 (1976).
15. L. E. Davis, in *Water-Soluble Resins* (R. L. Davidson and M. Sittig, Eds.) Reinhold, New York (1968) p. 216.
16. I. M. Klotz, G. P. Royer, and A. R. Sloniewsky, *Biochemistry* 8, 4752 (1969).
17. G. P. Royer and I. M. Klotz, *J. Am. Chem. Soc.* 91, 5885 (1969).
18. I. M. Klotz, G. P. Royer, and I. S. Scarpa, *Proc. Natl. Acad. Sci. USA* 68, 263 (1971).
19. H. C. Kiefer, W. Congdon, I. S. Scarpa, and I. M. Klotz, *Proc. Natl. Acad. Sci. USA* 69, 2155 (1972).
20. B. D. Batts, *J. Chem. Soc.* (B), 547 (1966).
21. W. J. Spetnagel and I. M. Klotz, *J. Am. Chem. Soc.* 98, 8199 (1976).
22. J. Suh, I. S. Scarpa, and I. M. Klotz, *J. Am. Chem. Soc.* 98, 7060 (1976).
23. R. H. Weatherhead, K. A. Stacey, and A. Williams, *J.C.S. Perkin II*, 800 (1978).
24. H. AC-Rawi, K. A. Stacey, R. H. Weatherhead and A. Williams, *J.C.S. Perkin II*, 663 (1978).
25. R. Breslow, H. Kohn, and B. Siegel, *Tetrahedron Lett.*, 1645 (1976).
26. W. E. Meyers and G. P. Royer, *J. Am. Chem. Soc.* 99, 6141 (1977).
27. D. Coleman and G. P. Royer, *J. Org. Chem.*, 45, 2268 (1980).

AUTHOR INDEX

AC-Rawi, H., 220
Adler, S. P., 180
Alberty, R. A., 69
Altman, C., 47, 87
Anantharamaiah, 159
Anderson, B. M., 140
Anfinsen, C. B., 38
Aoki, Y., 179
Aoyama, Y., 220
Archer, M. C., 204
Auld, D. S., 141

Bachovchin, W. W., 128, 141
Baker, B. R., 98, 112
Banaszak, L. J., 112
Banno, Y., 179
Baricos, W. H., 203
Barman, T., 159
Baronowsky, P., 112
Batts, B. D., 220
Bayley, H., 101, 112
Bender, M. L., 92, 112, 113, 122, 140, 208, 220
Benkovic, S. J., 113, 140, 152, 159
Bentley, R., 149, 159
Bergmeyer, H. U., 87
Berhauser, J., 141
Berliner, L. J., 187, 203
Bernfeld, P., 183, 203
Bernhard, S. A., 10, 13, 37, 140, 179
Biesecker, G., 87, 112
Birktoff, J. J., 141
Blake, C., 141
Blakeley, R. L., 112
Blout, E. R., 142

Blow, D. M., 112, 127, 141
Bovey, F. A., 142
Bowers, L. D., 204
Boyer, P. D., 140, 141, 179
Branden, C. I., 112
Breslow, R., 130, 141, 217, 220
Bridges, A. J., 112
Brocklehurst, K., 203
Brockman, H. L., 203
Broun, G., 204
Broun, R. S., 141
Brown, A. J., 87, 180
Brown, M. S., 180
Bruice, T. C., 113, 137, 140, 141, 142
Bryant, F. R., 152
Buckingham, D. A., 141
Buehner, M., 87
Buncel, E., 141
Bunting, P. S., 87, 192, 204
Burk, D., 87
Butler, L. G., 203

Cahn, R. S., 147, 159
Canady, W. J., 20, 38
Canfield, R., 134, 141
Carr, P. W., 204
Carrioulo, J., 114, 140
Cecere, F., 204
Chambers, R., 203
Chang, T. M. S., 203, 204
Changeux, J. P., 79, 162, 165, 166
Chibata, I., 197, 203, 204
Chipman, D. M., 142
Chock, P. B., 180
Clark, A., 159

AUTHOR INDEX

Cleland, W. W., 67, 69, 87
Clement, G. E., 140
Clowes, G. A., 220
Cohen, J. A., 112
Cohen, P., 179
Cohen, W., 203
Cohlberg, J. A., 179
Cohn, M., 110, 112
Coleman, D., 220
Coleman, J. E., 112
Colowick, S. P., 159
Colton, C. K., 204
Congdon, W., 220
Conn, E., 108
Conway, A., 168, 179
Cooney, C. L., 204
Cordes, E. H., 140
Cox, J. M., 140
Cramer, F., 220
Crestfield, A. M., 21, 38, 93, 111
Crook, E. M., 203
Crosby, J., 139, 142
Czarnieki, M. F., 220

Dalziel, K., 69
Danielson, B., 204
Darnall, D. W., 38
Davidson, R. L., 220
Davie, E. W., 180
Davis, L. E., 220
Delbaere, L. T., 141
d'Heck, H., 140
Dickerson, R. E., 37, 104, 111, 140
Dingwall, C., 157, 159
Dixon, M., 38
Dolphin, D., 141
Dowd, J. E., 87
Drenth, J., 141
Duntze, W., 179
Dymowski, J. J., 141

Eadie, G. S., 54, 87
Edsall, J. T., 37
Eisenberg, D., 37
Eisensach, J., 141
En-el, Z., 38
Engasser, J. M., 190, 203
Engleman, D. M., 159

Evans, D. R., 179
Everse, J., 141
Eyl, A. W., 37, 93, 111

Falderbaum, I., 141
Farnham, S., 17, 37
Feeney, R. E., 111
Ferdinand, W., 37
Feisht, A. R., 79, 129, 141, 157, 159, 169
Fife, T. H., 130, 141
Figueriedo, A. F. S., 12, 37
Filmer, D., 179
Finn, F. M., 38
Fischer, E., 153
Fleming, A., 133
Ford, G. C., 87, 112
Franzen, J. S., 17, 37
Frey, P. A., 152, 159
Friesen, H-J., 38
Fruton, J. S., 153, 159
Fujikawa, K., 180
Fulton, S. P., 204
Fung, C. H., 140
Fuse, N., 203

Gabel, D., 187, 203
Garatini, S., 203
Gardner, C. R., 204
Gazyola, C., 112
Geiss, I., 104
Gerhart, J. C., 179
Ginsburg, A., 179
Goodman, M., 141
Goldman, R., 192, 204
Goldstein, L., 184, 189, 203, 204
Graves, J. M. H., 159
Gruhn, W. B., 220
Guilbault, G. G., 204
Gulinelli, S., 203
Gupta, R. K., 140

Haber, E., 38
Haldane, J. B. S., 45
Hamilton, B. K., 204
Hamilton, G., 140
Hanes, C. S., 69
Hanson, K. R., 149, 159
Harris, J. I., 87, 112
Harrison, L. W., 141

AUTHOR INDEX

Hartley, B. S., 111, 141
Hartsuck, J. A., 140
Hassner, J. A., 159
Hatano, H., 154, 159
Hayashi, R., 159
Heilmeyer, L., 180
Heinrich, P. C., 180
Henri, V., 43, 87
Herschfield, R., 220
Hexter, C. S., 112
Hicks, G. P., 204
Hirschmann, 149, 159
Ho, H. T., 159
Hofmann, K., 38
Hofstee, B. H. J., 182, 203
Hol, W. G. J., 141
Holbrook, J. J., 141
Holzer, H., 172, 179, 180
Hornby, W. E., 192, 203
Horvath, C., 190, 203
Hruska, J. F., 38
Hsiao, H., 159
Hunkapillar, M. W., 141
Hutcheon, W. L. B., 141
Hymes, A. J., 38

Ichikawa, T., 159
Inagami, T., 37, 93, 111, 154, 159
Ingold, C. K., 147, 159
Inouye, K., 153, 159
Isowa, Y., 159

Jacob, F., 179
Jaffer, S., 141
James, M. N. G., 141
Jansonius, J. N., 141
Jenks, W. P., 37, 113, 114, 137, 140, 142, 155, 156, 159
Johnson, L. F., 33, 35, 38
Jones, J. B., 152, 159

Kahn, J., 203
Kaiser, E. T., 141
Kano, K., 220
Kantrowitz, E. R., 179
Kaplan, N. O., 141
Karube, I., 204
Karush, F., 38
Katchalski-Katzir, E., 187, 203, 204

Katunuma, N., 179
Kauzmann, W., 19, 37, 38
Kay, L. M., 37, 111, 140
Kean, C. E., 204
Kedem, D., 204
Kemp, A. S., 142
Kezdy, F. J., 112, 140, 203
Kida, M., 220
Kiefer, H. C., 214, 220
Kilby, B. A., 111
King, E. L., 47, 87
Kingden, H. S., 180
Kirby, A. J., 129, 141
Kirkuchi, J., 220
Kito, K., 179
Klapper, M. H., 37
Klotz, I. M., 17, 37, 38, 139, 142, 179, 213, 220
Knowles, J. R., 110, 112
Koch, G. L. E., 179
Koch-Schmidt, A. C., 203
Koeppe, R. E., 141
Kohn, H., 220
Kolkoek, R., 141
Kominami, E., 179
Komiyama, M., 220
Konecny, J., 186, 198, 203, 204
Koshland, D. E., Jr., 154, 159, 168, 179
Kraus, A., 133, 141
Krebs, E. G., 170, 179
Kuo, L. C., 141

Laidler, K. J., 87, 192
Langerman, N. R., 38
Larsson, P. O., 201, 204
Lauger, R. S., 204
Laursen, R., 112
Law, J. H., 38, 203
Lee, C., 141
Lee, W. W., 112
Leher, S. S., 142
Letsinger, R. L., 210, 220
Levin, Y., 187, 203
Levitt, M., 10, 37, 136, 142
Lewis, S. D., 37
Liberatore, F. A., 203
Lienhard, G. E., 135, 142
Lieso, K., 180
Liljas, A., 112, 141

Lilly, M. D., 203
Lineweaver, H., 87
Lipscomb, W. N., 110, 112, 128, 140, 179
Loeb, L. A., 112
Lotan, N., 141
Lowe, C. R., 38

McClure, D. E., 141
McDonald, R. C., 157, 159
Magni, G., 180
Makinen, M. W., 130, 131, 141
Malhotra, O. P., 179
Manecke, G., 184, 203
Mangum, J. K., 180
Mann, T., 109, 112
Mansson, J. D., 204
Marconi, W., 203, 204
Mares-Guia, M., 12, 37
Marikawa, Y., 201, 204
Markert, C. L., 141
Markley, J. L., 128, 141
Masek, Z., 180
Masters, C. J., 180
Mattes, S. L., 85, 87
Mattiason, B., 194, 204
Means, G. E., 111
Mecke, D., 172, 180
Medviczky, N., 37
Melamud, E., 140
Meloche, H. P., 112
Menten, M. L., 87
Meyers, W. E., 217, 220
Michaelis, L., 43, 87
Mildvan, A. S., 110, 112, 126, 140
Miles, J. L., 38
Millar, D. B., 141
Miller, S. T., 203
Miron, T., 38
Molhatra, O. P., 112
Moller, F., 141
Monod, J., 162, 165, 166, 179
Montalvo, J. G., 204
Moore, S., 21, 38, 91, 93, 111
Moras, D., 87, 112
Moravetz, H., 210, 220
Mori, T., 203
Morisi, F., 203, 204
Mosbach, K., 194, 204
Muir, G., 141

Muirhead, H., 140
Müller-Eberhard, H. J., 180
Mulvey, R. S., 179
Murachi, T., 154, 159
Murakami, Y., 220

Neat, K. E., 159
Nelson, G. H., 38
Nemethy, G., 179
Nicolson, G. L., 38
Nixon, N. E., 112
North, A., 141
Nozaki, Y., 38

O'Connell, E. L., 151, 159
Ogston, A. G., 148, 159
Ohmori, M., 159
Okahata, Y., 220
O'Leary, M. H., 85, 87
Olsen, K. W., 82
Ong, E. B., 112
Oosterbaan, R. A., 112
Osen, K. W., 112
Overberger, C. G., 211, 220
Owens, J. D., 159

Page, M. I., 137, 142
Pandit, U. K., 137, 141, 142
Pardee, A. C., 179
Pasteur, L., 146
Pastra-Landis, S. C., 179
Patt, S. L., 141
Paul, K., 142
Pauling, L., 16, 19, 37
Penna, G. D., 204
Perham, R. N., 112
Perutz, M. F., 87
Pestka, S., 179
Pfleiderer, G., 141
Phillips, D., 134, 141
Pigiet, V. P., Jr., 179
Plowman, K. M., 87
Porubcan, M. A., 128, 141
Prelog, V., 147, 159
Purich, D., 180

Quiocho, F. A., 112, 203

Rando, R. R., 112

AUTHOR INDEX

Rapperole, B., 204
Ray, W. J., 154, 159
Recsei, P. A., 109, 112
Reed, L. A., 38
Richard, J. P., 159
Richards, J. H., 141
Richards, K. E., 179
Riley, N. D., 112
Riggs, D. E., 87
Ringold, H. J., 159
Roberts, J. D., 128, 141
Robillard, G., 128, 141
Robinson, D. H., 38
Rose, I. A., 151, 159
Rosen, O. M., 179
Rosenberg, H., 37
Ross, L. O., 112
Rossman, M. G., 87, 106, 112, 141
Royer, G. P., 38, 203, 213, 217, 220
Rubin, C. S., 179

Sabesan, K. N., 87, 112
Saenger, W., 220
Salamone, J. C., 211, 220
Salmona, M., 203
Sanger, F., 91
Sarma, V., 141
Saronia, C., 203
Savereide, T. J., 210, 220
Scarpa, I., 142, 220
Schachman, H. K., 179
Schaefer, J., 112
Scheraga, H. A., 38
Schmir, G. L., 140
Schoellman, G., 112
Schutt, H., 180
Schwert, G. W., 141
Sebastian, J. F., 220
Secemski, I., 142
Segal, A., 180
Segel, H. J., 87
Segel, I. H., 40, 43, 62, 86, 87
Seydoux, F., 179
Shafer, J. A., 15, 37
Shaltiel, S., 38
Shapiro, B. M., 172, 180
Sharon, N., 142
Shaw, E., 99, 112
Shemin, D., 97

Shotton, D. M., 140
Shulman, R. G., 128, 141
Simon, H., 133, 141
Siegel, B., 220
Singer, S. J., 30, 38
Singh, A., 100
Sittig, M., 220
Sloan, A. I., 140
Sloan, D. L., 112
Sloniewsky, A. R., 142, 220
Smallcombe, S. H., 141
Snell, E. E., 109, 112
Sober, H. A., 5
Sonenberg, N., 112
Spatz, L., 30, 38
Spatz, H. C., 220
Spetnagel, W. J., 220
Squillacote, V. L., 130, 141
Srere, P. A., 195, 205
Stacey, K. A., 220
Stadtman, E. R., 172, 179, 180
Stein, W. H., 21, 91, 93, 111
Steindel, S. J., 141
Steinberg, I. Z., 187, 203
Steitz, T. A., 159
Strittmatter, P., 30, 38
Stroud, R. M., 37, 94, 111, 140, 141, 142
Studier, F. W., 38
Stumpf, P. K., 108
Suh, J., 142, 220
Sumner, J. B., 110, 112
Sunamoto, J., 220
Suzuki, S., 204
Swartz, W. E., 203
Sykes, B. D., 141

Tahara, T., 220
Tanford, C., 38
Tannenbaum, S. R., 204
Thierry, J. C., 87, 112
Thiessen, W. E., 141
Thomas, D., 204
Thomson, A., 112
Thornton, E. R., 100
Tong, E., 112
Tosa, T., 197, 203, 204
Trayser, K. A., 159
Turnquest, B. W., 140

Umbarger, H. E., 179
Updike, S. J., 202, 204
Urata, G., 179
Uy, R., 203

Vallee, B. L., 109, 110, 112, 128, 141
Van Adrichem, M., 112
Van Etten, R. L., 220
Voser, W., 198, 203, 204

Walker, J. E., 87, 112
Wan, J., 183, 203
Warren, S. G., 179
Warshal, A., 37, 142
Watsen, H. C., 140
Weatherhead, R. H., 216, 220
Weaver, J. C., 204
Webb, E. C., 38
Weber, G., 179, 180
Weber, K., 163, 179
Weetall, H. H., 184, 203
Weissbach, H., 179
Wells, M. A., 141
Wernick, D. L., 131, 141

Westheimer, F. H., 100, 112, 140
Wharton, C. M., 189, 203
Whitaker, D. R., 141
Whitesides, G. M., 204
Wiedemann, L. M. W., 33, 35, 38
Wilchek, M., 38, 112, 203
Wildnauer, R., 38
Wiley, D. C., 179
Wilkinson, G. N., 86, 87
Williams, A., 220
Williams, R. C., 179
Williams, R. J. P., 110, 112
Wingard, L. B., 203, 204
Woenckhaus, C., 141
Wonacott, A. J., 87, 112
Wong, J. T., 69
Wuff, K., 180
Wyman, J., 37, 165, 166

Yamamura, K., 141

Zahn, H., 38
Zamier, A., 112
Zernier, B., 110, 112

SUBJECT INDEX

α-effect, 124
Acetate decarboxylase, 125
Acetate kinase, 201
tetra-N-Acetylchitotetraose, 135
Acetyl choline, 12
Acetylcholinesterase, 12, 94
Acetylesterase, 198
Acetyl imidazole, 114
Acid-base, catalysis, 113
Activated complex, 42
Activation, free energy, $\Delta G^{\ddagger}$, 59
Activation energy, ΔE, 59
Active center, 9
Active site, 11
  constituents of, 92
  pyruvyl residue, 109
Active transport, 196
Acylase, immobilized, 197
Acyl imidazole, 118
Adenylate cyclase, 170, 171
Adenylate kinase, 201
Adenylating enzyme, 173
Adenylation, 174
Adenylation/deadenylation, 170
ADP ribosylation/de-ADP ribosylation, 170
Adrenalin, 170
Affinity chromatography, 32, 34
Affinity labeling, 98
Agarose, 184
Alanine, 4
Alcohol dehydrogenase, 110, 125, 133, 157
Aldolase, 96, 125

Alkaline phosphatase, 110
Allosteric enzymes, 162
Allosteric site, 9
L-Amino acid acylase, 197
Amino acids, table, 2
Aminoalkyl agarose, 32
δ-Aminolevulinic acid dehydratase, 97
6-Amino penicillanic acid, 198
c-AMP-dependent protein kinase (cAMP, Prk), 170, 171
Amplification cascade, 172
Anchimeric assistance, 129
Anisole chlorination, 217
"Anticooperative," 168
Apoenzyme, 105
Apolar (hydrophobic) bond, 11, 19, 24, 25
Arginine, 3
Arrhenius law, 58
Arrhenius plot, 59, 131
Arsanilloazocarboxypeptidase, 128
Arylazides, 101
Ascorbic acid, 8
Asparagine, 2
Aspartame(aspartyl-phenylalanine methyl ester), 144
Aspartate aminotransferase, 102
Aspartate transcarbamylase, 24, 110, 163, 164
Aspartic acid, 3
Asp-His-Ser catalytic triad, 103
Assays: continuous, one-point, 84
  coupled, 85
ATCase (aspartate transcarbamylase), 24, 110, 163, 164

SUBJECT INDEX

Atomic absorption spectroscopy, 110
ATP regeneration, 201
4-Azidocinnamoyl group, 101

β-structure, 24
Basicity, 122
Benzamidine, 12, 93
N-Benzoyl-L-arginine ethyl ester, 12
N-Benzoyl-L-tyrosine ethyl ester, 99
Biotin, 7, 107, 709
Bireactant systems, 69
Blood clotting, 175, 178
Boltzmann's constant, 43
Brønsted β, 122
Brønsted equation, 122
Brønsted plot, 123
"Bucket brigade" enzymes, 28

Cacoonase, 26, 32
Carbenes, 100, 101
Carbobenzoxy group, 218
Carbodiimide, water soluble, 93
Carbonyl group polarization, 125
Carboxypeptidase A, 110, 126, 128-131
Carboxypeptidase B, 110
Carboxypeptidase Y, 157, 199
Cardiolipin, 30
Catalytic domains (dehydrogenases), 106
Catalytic subunits, 163
Catalytic triad, serine proteases, 127
Cephalosporins, 198
Charge-relay system, 103, 127
Chelate effect, 11
Chemical modification of enzymes, 93
  enzyme catalyzed, 172, 178
Chiral center, 146
Chitin synthase, 178
O-(trans-p-Chlorocinnamoyl)-L-β-phenyl-
  lactate, 130, 131
Cholesterol, 29
Choline, 12
α-Chymotrypsin, 20, 94, 99, 101, 103, 118
Chymotrypsin C, 175
Chymotrypsinogen, 176
Cleland's rules, 80, 84, 89
Coenzyme, 1,6, 104
Coenzyme A, 8
Cofactor, 104, 200
Co-immobilized enzymes, 194, 195

Competitive inhibition, 61
Compound S, 198
Cosubstrate, 104
Coupled assay, 32
Covalent inhibition, 67
Covalent intermediates, 96
Covalent structure, 20
Cycloamyloses, 206
Cyclodextrins, 207, 208, 217
Cycloheptaamylose, 207
Cyclohexaamylose, 207
Cyclooctaamylose, 207
Cysteine, 2
Cytochrome b, 30
Cytochrome b_5 reductase, 30

Dead-end inhibitor, 79, 84
Deadenylation, 173, 174
Decarboxylation, 215
$\Delta^{1,2}$-Dehydrogenase, 199
Delocalized charge, stabilization of, 216
Denaturation, 9, 20, 24
Desaturase, 31
Desensitization, 9
Destabilization, 135, 138-140
Deuridylylation, 174
Deuterium isotope effect, 121, 130, 212
Diamines, pK_a's, 14
Dielectric constant, 10, 11
Differential labeling, 94
Diffusional effects, 186, 190
Diffusional limitations, 181
  external, 182, 191
  internal, 182, 192
Dihydrofolate reductase, 33, 35
Dihydrolipoyl dehydrogenase, 27
Dihydrolipoyl transacetylase, 27
Dihydroxyacetone phosphate, 97
Diisopropylphosphofluoridate (DFP), 94
Dinucleotide-binding domains (dehydro-
  genases), 106
Dipole-dipole interactions, 24
Distortion, 138
Distribution equations, 49
Disulfide bridge, 25
Dixon plot, 64
DNA polymerase, 110, 126
Double reciprocal plot, 86
"Double-sieve" editing model, 157, 158

SUBJECT INDEX

Eadie-Hofstee plot, 55, 86
EDTA, 110
Effectiveness factor, 192, 193
Elastase, 103
Electron "sink," 124
Electrophilic catalysis, 113, 124-127, 209
Electrostatic bond, 11
Ellman's reagent, 216
Enantiomeric specificity, 152
cis-Enediol, 151
Energy of activation, 59, 60
"Entatic" centers, 136
　　state, 110
Enterokinase, 178
Enthalpy of activation, $\Delta H^{\ddagger}$, 59, 60
Entropy of activation, $\Delta S^{\ddagger}$, 59, 60
Entropy loss, 137
Enzyme: electrode, 202
　　extracellular, 26
　　immobilization, 182, 183
　　intracellular, 26
　　localization, 26
　　membrane, 196
　　membrane-bound, 28
　　in organic synthesis, 198
Enzyme-substrate complex, 10
Epinephrine, 170
Extraction model, 20
Extrinsic pathway (blood clotting), 175, 178

Factor X, 175
Factor Xa, 177
FAD, 6
Faraday constant, 188
FARCE, 137
Feedback inhibition, 162
Ferrocinnamic acid, 206
Fibrin, 177
First-order rate equation, 40
Flavine adenine dinucleotide, 6
Fluid mosaic model, 30
β-2-Furylacryloyl phosphate, 96

β-Galactosidase, 198
General acid catalysis, 117
General base catalysis, 115
Glass: porous beads, 184
　　activation of, 186

Glucose-6-phosphate dehydrogenase, 133
Glutamate dehydrogenase, 98, 133
Glutamic acid, 3, 120
Glutamine, 3
Glutamine synthetase, 172, 174
Glyceraldehyde-3-phosphate dehydrogenase, 95, 105, 106
Glycinamide, 93
Glycine, 2
Glycogen breakdown, 171
Glycogen phosphorylase, 170, 171
Glycolic acid, 101
Glycosyloxocarbonium ion, 135

Half-site reactivity, 168
Half-time, 40
α-Helix, 21, 23
Helix content of some globular proteins, 24
Henri-Michaelis-Menten equation, 43
　　linear forms, 53
Heterotropic activation, 164
　　effector, 165
　　inhibition, 164
Hexokinase, 154, 155, 157
Hill coefficient, 167, 168
Hill equation, 168
Hill plot, 168
Histidine, 3, 120
Histidine decarboxylase, 109
Holoenzyme, 105
Homomorphic groups, 140
Homotropic activation, 164
Hormonal control, 171
Hydrazine, 124
Hydride ion transfer, 133
Hydrogen bond, 16, 25
　　in ES complex, 18
Hydrogen bond acceptor and donor, 16
Hydrogenolysis, 219
Hydrolase, 143
Hydrophobic anchor, 30
Hydroxamate group, 212
α-Hydroxyethyl-thiamine pyrophosphate, 27
Hydroxylamine, 124
11-β-Hydroxylase, 199

Induced-fit theory, 13, 153, 154, 155
Initial velocity, 45

SUBJECT INDEX

Integral proteins, 30
Intrinsic pathway (blood clotting), 175, 178
Iodoacetic acid, 93
Ion pairs, 25
Isoamyl alcohol, 12
Isocitrate dehydrogenase, 133
Isoleucine, 5
Isomerase, 144
Isotope exchange, 79, 84

K_{cat}, 53
K_m, Michaelis constant, 44, 53
King-Altman method, 47, 71
Kings patterns, 49
KNF (sequential) model, 169
Krebs cycle, 196

Lactate dehydrogenase, 105, 106, 132-133
Lactose intolerance, 198
LADH, liver alcohol dehydrogenase, 105, 106
Leaving group, catalysis by, 121
Leucine, 4
Ligase, 144
Lineweaver-Burk plot, 54, 86
Lipoic acid, 107, 109
Lipoyllysine, 28
Lock and Key hypothesis, 153
Lyase, 144
Lysine, 3, 120
Lysozyme, 104, 128-133
α-Lytic protease, 103, 128
Lactate dehydrogenase, 105, 106, 132-133
Lactose intolerance, 198

Macrocycles, 206
Malate dehydrogenase, s-MDH, 105, 106
Malic enzyme, 125
Mandelate racemase, 125
Medium effects, 219
Membrane-bound enzymes, 181
Metal ions, 1, 109
Metallo-enzymes, 125
Methane-stannane series, 18
Methionine, 5
N-methylacetamide, 17
Methyl orange, 213
Michaelis complex, 43
Michealis complex, (K_m), 43

integrated, 88
Microencapsulation, 183
Microenvironmental effects, 186, 187
"Mischarging" t-RNA, 158
Mixed inhibition, 67
Molecularity, 40
Multienzyme complexes, 26
MWC (symmetry model), 165, 166, 169

NADH/NAD⁺, 6, 85
NAD-linked dehydrogenases, 105
(NAG-NAM)$_n$, 134, 135
Negative cooperativity, 168, 169
Nicotinamide adenine dinucleotide, 6
Nicotinamide ring, 133
Nitrenes, 100, 101
Nitrobenzisoxazole carboxylic acid, 139
4-Nitrocatechol sulfate, 118
p-Nitrophenyl diazoacetate, 100
p-Nitrophenyl trifluoroacetate, 218
^{15}N-NMR spectroscopy, 128
Noncompetitive inhibition, 66, 67
Nucleophilic catalysis, 113, 118-124

Oligomers, 24
Orbital steering, 137
Order (reaction), 40
Organic synthesis with enzymes, 200
Orientation, 136
Oxidoreductase, 143

Palladium-PEI-catalyst, 218
Papain, 15
Parabolic replot, 79
Paracyclophanes, 209
Penicillanic acid, 6-amino, 198
Penicillin G, 198
Pepsin, 153, 155
Pepsinogen-pepsin, 176
Peptide synthesis, 199
Peptidyltransferase, 101
Peripheral proteins, 30
Perspective formula, 146
1,10-Phenanthroline, 110
Phe-t-RNA, 102
Phenylalanine, 5
Phosphate binding protein, 13
Phosphoglycerides, 29
Phosphoglucomutase, 126, 153, 155

SUBJECT INDEX

Phosphorylase a, 170
Phosphorylase b, 170
Phosphorylase kinase, 170
Phosphorylase kinase kinase, 170
Phosphorylase phosphatase, 170
Phosphorylation/dephosphorylation, 170
Photoaffinity labeling, 100
pH-rate profile, 56, 181
pH-stat, 85
Ping pong pathway, 69
pK_a, 13, 122
Plank's constant, 43
β-Pleated sheet, 23, 25
Poly(ethylenimine), 184, 212, 213
Polymer-"ghosts," 217
Poly(U), 102
Poly(4,[5]-vinylimidazole), 21
Poly(4-vinyl pyridine), 211
Poly(vinylsulfonate), 210
Porphobilinogen, 97
Prednisolone, 198
Primary structure, 20
Peptide backbone, 22
Procarboxypeptidase A, 175
Prochiral center, 148, 149
Product inhibition, 19, 80
  patterns, 81, 83
Proenzymes, 161
Proline, 5
Pronase, 32
Propinquity, 137
pro-S, 149
pro-R, 149
Prosthetic group, 104
Protein conformation, 10
  purification, 33
Proteus B, *S. griseus*, 103
Protomers, 24
Proton transfer rate, 117, 118
Proximity, 136, 137
Pseudosubstrates, 94
Pyridoxal phosphate, 7, 96, 108, 124
Pyruvate carboxylase, 125
Pyruvate decarboxylase complex, 27
Pyruvate kinase, 126
Pyruvyl residue, 109
Purification table, 36

Quaternary structure, 24

Random-rapid equilibrium mechanism, 69, 70, 78
  double reciprocal plot, 76
Rate constant, 40
Rate enhancement, 136-140
Reaction order, 40
  and rate enhancement, 137
Reciprocal plot, 86
re-face, 150
Regiospecificity, 144
Regulatory subunits, 163
re-si face, 151
re-re face, 151
Reversible one-substrate reactions, 52
Ribonuclease, 9, 21, 93
Ribulose diphosphate cardoxylase, 125
Rotomer distribution, 137
R,S descriptions, 147

Salicylic acid, 98
  4-(iodoacetamido), 98
Salt bridges, 11, 24
Schiff base, 96, 97, 106, 125
Secondary structure, 23
Second-order rate equation, 41
Second-sphere complexes, 126
Serine, 2
Serine hydroxyl, 120
Serine proteases, 127-128
Sequential model, 169
si-face, 150
S_o, initial substrate concentration, 44
Small-angle x-ray scattering, 157
Sodium borohydride, 96, 97, 107, 125
Solvation, 139
Specific acid catalysis, 114
Specific base catalysis, 114
Specificity, 143
  catalytic, 153
  structural, 144
  stereochemical, 146
  stereoheterotopic, 148
  reaction, 143
Steady-state approach, 45
Stereoheterotopic, 150
Stereopopulation control, 137
Steroid synthesis, 198
Strain, 138
Substrate, 1

SUBJECT INDEX

binding energy, 19
Subtilisn, 103
Suicide substrate, 102
"Synzymes," 205

Tertiary structures, 24
Tetrahydrofolic acid, 7
Theorell-Chance pathway, 69, 70, 77, 79
  double reciprocal plot, 76
Thermal denaturation, 57, 60
Thermal stability, 61
Thermolysine, 110, 144
Thiamine pyrophosphate, 7
Thiamine-pyruvate adduct, 139
Threonine, 2
Threonine deaminase, 162
Thrombine, 177
TLCK, 99
Togetherness, 137
T_{opt}, optimum temperature, 58
TPCK, 99
TPP-thiamine pyrophosphate, 28
Transamidase, 177
Transaminases, 96
Transcarboxylase, 125
Transferase, 143
Transition state theory, 42
Translocases, 28
Transplanar peptide bond, 22
Trypsin, 11, 93, 103, 155
  inhibitor, 175
Trypsinogen, 175

Trystophan, 4
Tyrosine, 4
  hydroxyl group, 120
Tyrosyl-t-RNA synthetase, 168, 169

Uncompetitive inhibition, 65
Unit of enzyme activity, 36
Unstirred layer, 182, 190
Urea, 20, 36
Urease, 110, 136
Uridylyl-removing enzyme (UR enzyme), 173
Uridylylation, 174
Uridylyl enzyme, 173
Utase, 173

Valine, 4
Valine chloromethylketone, 100
Valyl-t-RNA synthetase, 100
van der Waals' forces, 9, 11, 18, 24, 25
van der Waals' radius, 19
van't Hoff equation, 60

Wrong-way binding model, 153, 154, 155, 156

X-ray crystallography, 16, 91, 103-104

Yeast alcohol dehydrogenase, 20

Zymogens, 174, 176